belfast
met

Millfield L

Sustainable Thermal Storage Systems

About the Author

Lucas B. Hyman, P.E., LEED AP, is President of Goss Engineering, Inc. A professional mechanical engineer with nearly 30 years' experience, he has planned and designed thermal energy storage systems, central heating and cooling plants, cogeneration plants, and other mechanical systems. Mr. Hyman has participated as a mechanical engineer member in value engineering project reviews, conducted and led forensic engineering studies, and served as an expert witness. He is co-editor of *Sustainable On-Site CHP Systems: Design, Construction, and Operations*, published by McGraw-Hill.

Sustainable Thermal Storage Systems

Planning, Design, and Operations

Lucas B. Hyman, P.E., LEED AP

New York   Chicago   San Francisco
Lisbon   London   Madrid   Mexico City
Milan   New Delhi   San Juan
Seoul   Singapore   Sydney   Toronto

The McGraw·Hill Companies

Sustainable Thermal Storage Systems: Planning, Design, and Operations

1 2 3 4 5 6 7 8 9 0   DOC/DOC   1 7 6 5 4 3 2 1

ISBN 978-0-07-175297-8
MHID 0-07-175297-8

This book is printed on acid-free paper.

Sponsoring Editor
Joy Evangeline Bramble

Editing Supervisor
Stephen M. Smith

Production Supervisor
Richard C. Ruzycka

Acquisitions Coordinator
Alexis Richard

Project Manager
Aloysius Raj,
Newgen Publishing
and Data Services

Copy Editor
Priyadharshini
Dhanagopal

Proofreader
Helen Mules

Indexer
Doreen McLaughlin

Art Director, Cover
Jeff Weeks

Composition
Newgen Publishing
and Data Services

Contents

Preface

With ever-more-constrained electric grids and higher energy costs, utility companies and facility owners are continuing to look at thermal storage as a proven method to reduce energy costs and reduce on-peak demand.

Thermal storage is a way of producing a thermal output at one point in time and using that thermal output during a different period of time. Using thermal storage can result in the following:

- Reduced on-peak electric demand
- Lower energy costs
- Smaller required equipment capacity to meet peak thermal demand
- Lower required capital costs
- Lower life-cycle costs
- Improved operational flexibility
- Less pollution

This book provides proven techniques for reducing energy and capital costs with thermal storage systems and provides architects, engineers, contractors, and facility managers with a primer on the sustainable planning, design, and operation of thermal storage systems. The book offers readers a number of key benefits including a good understanding of the following:

- What thermal storage systems are and how they operate
- What the requirements to make a thermal storage system sustainable are
- How to determine if a thermal storage system is right for your building or facility
- How to conduct a facility evaluation or feasibility study
- The advantages and disadvantages of various thermal storage systems
- How the nature of the load profiles drives a thermal storage project
- The relationship between loads, equipment choices, and system selection
- The issues affecting thermal storage location

- How to conduct estimated energy use and cost analysis
- The principles of economic analysis
- Start-up, testing, and commissioning requirements

Learn how to plan, design, and operate a sustainable thermal storage system using the detailed information provided in this practical guide!

Acknowledgments

I would like to thank the following individuals for their help in writing this book and for their support: Joel Goss, PE, retired; John Andrepont, PE, Principal, Cool Solutions; Brian Silvetti, PE, Vice President of Engineering, Calmac Corporation; Gennady Ziskind, D.Sc., Associate Professor and Head, Department of Mechanical Engineering, Ben-Gurion University of the Negev, Israel; Hank Bagheri, Ph.D., PE, Principal Engineer, Goss Engineering; Ed Mendieta, QCxP, LEED AP, CEM, Sustainability Consultant, TRC; Shaw Gentry, EIT, Goss Engineering; Ana Perez, AutoCAD Manager, Goss Engineering; Guy Frankenfield, PE, Thermal Storage Manager, Natgun; Justin Anderson, Project Manager, VSL; and David Bain, Business Development, DYK. I would also like to thank Ed Knipe, deceased, who first taught me about thermal storage, and my wife, Meggie, for her support and patience while I undertook this writing effort.

Lucas B. Hyman, P.E., LEED AP

CHAPTER 1

Overview

Introduction

Modern thermal storage is a proven technology with numerous benefits for multiple stakeholders. In fact, thermal storage has been saving money for knowledgeable facility managers and building owners for several decades. Utility companies have also benefited from the implementation of thermal storage over the years. The savings to utility companies, by eliminating the need for additional power generation and transmission construction simply to meet on-peak electric load, is so valuable to utilities that many utilities have provided financial incentives to facility owners to reduce the first cost of thermal storage installation. In addition, utilities often offer electric rates that provide incentive to transfer consumption of power from on-peak to off-peak periods, when utility-wide electric demand is much lower.

This book provides proven techniques for reducing energy use and cost and even potentially reducing capital costs with the use of thermal energy storage. It also provides a practical guide to the planning, design, basic construction, and operation of sustainable thermal storage systems.

There are several different types of energy storage methods including mechanical energy storage, chemical energy storage, magnetic field energy storage (an up-and-coming field), and thermal energy storage. Each energy storage method has numerous storage technologies such as, for example, mechanical energy storage—pumped hydro, compressed air, and flywheels; chemical energy storage—fuel oil storage; and many different types of battery storage technologies. However, this book discusses thermal storage, which relates to storing sensible or latent thermal energy; it focuses primarily on common applications of thermal storage in use today for building or facility heating and cooling.

The addition or removal of sensible energy causes a change in a material or substance's temperature, whereas the addition or removal of latent energy causes a material or substance to change phase (e.g., turn from water to ice). These processes are discussed in detail in later chapters.

Thermal storage is a low-cost form of energy storage that is often highly efficient. In fact, some thermal storage systems provide heating or cooling with less input energy than a nonstorage conventional production at time of use system. When combined with less expensive energy input, thermal storage provides a powerful "green" tool for the building or facility owner.

There are both passive and active thermal storage systems. Active thermal storage systems use mechanical production and transportation heating or cooling equipment

Figure 1-1   Aboveground stratified chilled water thermal storage tank under construction.

and prime movers such as fans or pumps with ducts or piping; passive thermal storage systems do not. An example of passive thermal storage is the sun shining on a high mass wall. However, this book focuses primarily on active thermal storage systems.

At its most basic, thermal storage provides methods to separate the time when heating or cooling is produced or generated from the time when that heating or cooling is needed or used. Thermal storage is sometimes defined as a way of producing an energy sink or source at one point in time and using it at another point in time. Thermal storage, therefore, provides methods and systems that allow storage of either cooling or heating produced at one period in time for later use at another period in time. For example, cooling can be produced at night in a more efficient manner than during the day at much less costly energy rates for use during the next day. Likewise, for example, solar heating can be produced during the peak sun output during the day and stored for space heating or domestic hot water use later in the evening or during morning warm-up. Cooling or heating can even be stored and used over our seasonal periods, for example, storing cooling in winter for use during summer, or storing heat in summer for use during winter.

Air-conditioning is often the largest single contributor to a building's energy costs, and when combined with other building energy use, the time of peak cooling energy needs produces the highest building electrical demand by far. That peak electrical demand is not only costly to the building owner in purchased energy, but it also creates challenges for the electric utility serving the building or plant. The utility may need to own and operate peak electric production and transportation equipment that is needed only for a few hours a year. To the utility that means a very costly plant investment for the revenue received.

Thermal storage is also sometimes called thermal energy storage or TES for short. Cooling thermal storage is sometimes called "cool storage" or "cooling storage." These terms are shortened versions of the original terms but understood by designers and users of such system even though not technically accurate, as it is not technically possible to store "cool." What the term means is to store an energy sink or a place that accepts heat removed from a building or structure.

As discussed further in this chapter, there are many benefits to the use of thermal storage including the following:

- Reduced utility-bill peak demand
- Improved facility plant efficiencies
- Reduced facility energy consumption
- Lower facility energy costs
- More flexible plant operations and added backup capacity
- Smaller equipment requirements
- Lower maintenance costs
- Potential lower capital costs
- Greater economic benefits than the business-as-usual case

Adding thermal storage can make the existing production equipment serve a far greater load than the installed equipment capacity. When thermal storage is combined with measures to increase the temperature rise or drop (i.e., delta-T) in existing air or water systems, the thermal storage solution often means greater loads can be served with the same pipes and pumps or fans and ducts.

Society also benefits with the installation of thermal storage due to less pollution as a result of more efficient facility and utility operations as well as less polluting utility operations at night (depending upon the fuel–power mix used by the utility to produce electric power during thermal storage charging).

As discussed in Chap. 2, thermal storage technology can be and has been used by a variety of facilities and buildings including residential, commercial, and institutional for both the public and private sectors.

One typical highly desirable condition to use thermal storage is to have a building(s) or facility where the loads are cyclical or otherwise variable over time. The time period is often daily but can be weekly, seasonal, or even short term, for example, 10 to 20 minutes for emergency cooling. If the facility or building has a very flat load profile, thermal storage may not be suitable for load shifting, unless there is a significant difference in energy costs at one part of the day over other periods of the day. Another case for thermal storage with a constant load is to serve as an emergency backup in case of loss of heating or cooling production equipment. However, most facilities or buildings have a cyclical load profile.

Thermal storage systems are often most economical in locations that have time-dependent energy costs. For example, many utility companies charge more for on-peak power consumption (power used during the day) than for off-peak power consumption (power used during the night). In addition to energy consumption charges, most utilities also charge for peak power use. Thermal storage offers a powerful tool to reduce peak power demand. Some utilities provide economic incentives such as super off-peak pricing to encourage off-peak cooling production as well as utility rebates for the installation of thermal storage. One recent design will receive more than $4 million from the local utility company to assist with their thermal storage installation. Another client buys power from the utility, as well as generating much of their own electric power on site, and dispatches thermal storage to store produced chilled water using very low-price power. The serving utility has a large nuclear plant base and during some nights actually pays the project to use electricity. At other times, the combined real-time-pricing demand costs for electricity may approach $1000 per megawatt (MW). During those expensive utility time periods, the stored chilled water is used to meet campus cooling needs. The use of thermal storage with smart meters and real-time-pricing offers the opportunity for enhanced energy cost savings through specifically targeted peak demand reduction in those hours with the highest overall energy cost rates.

Thermal storage is often the most economically attractive:

1. When new facilities or buildings are constructed
2. When new facility expansions are planned
3. When old plant equipment needs to be replaced with new equipment

Life-cycle cost analysis often shows that employing thermal storage is the lowest life-cycle cost option among the many alternatives analyzed.

Thermal storage can also provide important benefits to utility electric generating and combined heat and power (CHP) plants when used for combustion turbine inlet-air cooling (CTIC).

There are a number of materials that can be used as a storage medium in thermal storage systems. The most common material is water, either as a liquid or as a solid. The most typical water thermal storage systems include stratified chilled water thermal storage systems, different types of ice storage systems for cooling, and hot water storage systems for heating. As discussed in this book, each type of thermal storage material and thermal storage system has its own thermal storage, equipment, and system requirements. Each system also has its own advantages and disadvantages. The past two to three decades have refined the thermal storage systems commonly used so that over time many of the thermal storage systems that were once tried and proved less successful are no longer used.

Thermal storage is becoming ever more important as energy costs rise and capital resources are ever scarcer. In addition, thermal storage is an important piece of the "green" solution puzzle to successfully incorporate renewable energy resources such as solar power and wind power into the power grid.

Benefits of Thermal Storage

As outlined above, there are many benefits to the use of thermal storage. The following paragraphs provide an overview of some of the many benefits of thermal storage.

Reduced Utility-Bill Peak Demand

As will be shown and explained in this book, the use of thermal storage can allow a facility or building to reduce or virtually eliminate the on-peak electrical demand associated with facility cooling production. Demand is an engineering name for the instantaneous rate of consumption of the item being measured. Examples include flow, cooling load, heating load, or most common electric power. Utility companies, in particular, measure a facility's electric power use both by the amount of total electrical energy consumption (often measured in kilowatthours, kWh) over a particular time period (e.g., monthly) and also by the peak power consumption rate, that is, the maximum demand (often measured in kilowatts, kW) during certain utility rate time periods; the utility company measures, records, and charges for the maximum peak power demand during each utility rate time period. In deregulated or re-regulated utility environments, there may be real-time pricing (RTP) or spot market rates, which will also reflect time-of-use variations in electricity costs.

Fortunately, the facility on-peak (and even mid-peak) power demand, typically, the most expensive utility rate period, can be reduced through the use of thermal storage. For example, as shown in Chap. 7, and known as full storage, cooling production can occur at night (with subsequent recorded energy consumption and demand costs accruing to the facility during that nighttime period), with only the thermal storage system operating the next day, during the peak cooling load period, to supply all of the facility's cooling needs with all of the cooling production equipment turned off (except for chilled water distribution pumps and facility air-handling units, which supply conditioned air to building occupants, and are, of course, turned on whenever a facility is occupied, whatever the cooling or heating method employed). As cooling production equipment is turned off during the on-peak period in this example, no electric power is, of course, required to operate the cooling production equipment, and the facility electric power demand is, therefore, reduced accordingly. Note that the facility cooling or heating load or demand itself is not reduced, just the cooling production equipment demand required to meet the facility load during the on-peak period, as operations required to meet the facility cooling load are modified, and, as will be seen, smaller production equipment can be used to meet the facilities cooling load.

Improved Facility Plant Efficiencies

As noted above, with cool thermal storage, cooling production equipment (e.g., chillers) are often operated more at night, when air temperatures (dry-bulb and wet-bulb) are typically lower than during the day. The amount of work and hence energy required by cooling production equipment per unit of cooling produced is, in part, a function of the condensing temperature, which is a function of the outside air temperature. In fact, for installed cooling systems, at any given time, the condensing temperature is often a major factor in cooling system efficiency performance, since many of the other efficiency factors are fixed for an installed cooling system. Lowering the condensing temperature can improve cooling production equipment energy efficiencies by up to approximately 1.5 percent for each degree Fahrenheit that the condensing temperature is lowered below the design condensing temperature.

Another contributor to improved chiller efficiency, with thermal storage systems, is that chiller equipment runs at full load most of the time, resulting in better overall plant kW-per-ton performance than when chiller equipment runs at low part load. Further, many ancillary components such as condenser pumps, cooling towers and similar appurtenances are based on full chiller load and do not unload linearly (if at all) with cooling load reductions. Therefore, operating mechanical equipment at or close to full load is generally more efficient than part load operations, and plant fixed parasitic losses are proportionally minimized at full load. Additionally, a fair amount of mechanical equipment often operates more inefficiently at low part load than at full load. As a result, conventional nonstorage chiller plants often have very poor annualized performance due to excessive part load operations. However, there are some all-variable-speed systems that do try to minimize power consumption at part load by adjusting individual equipment operating speeds.

Reduced Energy Consumption

As suggested, operating cooling production equipment more fully loaded at night, when dry-bulb and wet-bulb air temperatures are lower, often results in higher facility cooling plant energy efficiency. This is due, in part, to lower condensing temperatures at night but is also due to greater efficiency of chiller operating at design point. Another factor is that when pumps and cooling towers are operating with a full design temperature difference, the energy consumption per ton of cooling produced is less. Many nonstorage conventional plants, which must operate most of the year at part load, can use more energy in pumping energy than in cooling production equipment such as chillers or ice-makers. Another factor is that a well-designed thermal storage often uses much higher delta-T and, therefore, less chilled water flow than conventional systems, which must match delta-T to the chiller. Most thermal storage systems use variable flow, and this also reduces pump energy use during part-load operations. Therefore, a well-planned, designed, and constructed thermal storage plant has lower overall cooling plant energy consumption than conventionally operated nonstorage cooling plants.

Lower Energy Costs

Lower facility electric demand results in lower utility demand costs, which are often a significant portion of the lower energy costs experienced by facilities with installed thermal storage systems. However, reduced energy consumption as well as shifting production from more expensive utility energy rate periods to less expensive utility energy rate periods also provide for lower facility energy costs. Based on experience, annual energy cost savings can be as much as 20 percent (and sometimes even more) depending upon the thermal storage system installed. Installations with real-time-priced power have shown even greater savings.

More Flexible Facility Plant Operations

Thermal storage provides for more flexible plant operations. The thermal storage system can act as a "buffer," and the rate of discharge from a thermal storage system is typically variable from near zero discharge up to a maximum discharge flow rate as designed. Thermal storage systems are often designed for facility peak load conditions, and, therefore, during non-peak load conditions (which are most of the hours of the year), plant operations can easily be adjusted so that, for example, more cooling production occurs during off-peak hours with less cooling production occurring during mid-peak hours and no cooling production occurring during on-peak hours (even if the thermal storage was designed as a partial storage system on the peak day). Further, during lower load periods of the year, it may likely be possible, for example, for a cooling production unit (e.g., a chiller) to be out of commission and for the cooling plant using thermal storage still to be able to meet the facility's cooling load. In addition, during the case of an unforeseen plant shutdown, it may still be possible to serve critical loads for a period of time from thermal storage (if the thermal storage was charged or partially charged when the unforeseen plant shutdown occurred).

Added Backup Capacity

A daily thermal storage system is typically sized for the peak design day load of the facility or building. On non-peak load days (which are the majority of the days throughout the year), the facility or building load can be met with less equipment capacity. Since less than full cooling generation equipment capacity is required during periods with lower loads, the additional generation equipment available is in a sense added backup capacity. In some cases, one or two chillers or similar equipment units can be nonoperational (e.g., down for maintenance), and the facility plant can still meet the facility load. In a real-life example, a chilled water storage system allowed for emergency cooling when the entire chilled water plant was down, by dumping ice into the storage tank.

Smaller Facility Equipment Requirements

As shown in Chap. 7, using thermal storage, it is possible to meet a facility's peak heating or cooling load with installed equipment smaller than the facility's peak load. For example, a thermal storage system can be sized so that on a peak day, the cooling production equipment operates at full load for the entire 24 hours (known as partial storage), and, therefore, the cooling production equipment size required with partial thermal storage to meet the facilities load is equal to the average load on the peak day (plus any desired spare capacity). With a conventional cooling plant (no thermal storage), for example, the cooling production equipment needs to be sized equal to the facility's peak cooling load (plus any desired spare capacity). With a partial thermal storage system, depending upon the facility load profile, it is often possible to use cooling equipment only half the size of the facility's peak load and still meet the facility peak load.

Lower Facility Maintenance Costs

Maintenance costs are, in part, proportional to the number and capacity of the installed equipment; and, therefore, less equipment capacity requires less maintenance and results in lower maintenance costs. Additionally, operating mechanical equipment most of the time at full load with less starts and stops reduces maintenance requirements and subsequent maintenance costs. Therefore, some thermal storage systems can require less maintenance costs per unit of thermal energy delivered.

Potential Lower Capital Costs

A properly planned and designed thermal storage system often provides lower capital costs in addition to lower energy use and lower energy costs. This is due to smaller equipment requirements and smaller related infrastructure requirements, which is possible even with the added cost of the thermal storage tank. For example, a thermal storage system may only require 1000 tons of cooling equipment versus 2000 to 3000 tons of cooling equipment required for a conventional cooling plant with no thermal storage. Not only does the cooling production equipment cost much less, but the electrical and building infrastructure costs to support the cooling equipment will also be lower, potentially more than offsetting the capital cost of the thermal storage system, and, thus, producing a net capital cost savings versus a nonstorage system. Depending on the tank location and space demands, it is possible that there is also land cost savings.

Economic Benefits

Ongoing lower utility energy costs as well as potentially lower construction capital costs obtained when a facility uses thermal storage compared to a conventional nonstorage facility plant provide initial and ongoing life cycle direct economic benefits to a facility or building owner. Therefore, thermal storage systems can provide initial capital cost savings and ongoing energy cost savings for a combined economic benefit.

Environmental Benefits

As many utilities presently use fossil fuels to generate power, and the burning of fossil fuels harms the environment due to stack emissions (NOx, SOx, CO, particulates, etc.) as well as CO_2 emissions [considered by the majority of scientists today (2011) as a likely factor in global warming], using less energy with a well-designed and constructed thermal storage system typically results in less pollution. Additionally, at night, utilities typically operate their most efficient or most cost effective power production equipment to meet their base nighttime electric load. This base electric load often includes nuclear, hydroelectric, wind, and other green-energy power production systems as well as coal-fired systems. In contrast, many peak energy plants burn fossil fuel such as natural gas or oil, which typically increases the carbon burned over the base-load power plant (except for base-load coal power plants). Utility plants are also often more efficient at night than during the day, also due to the colder nighttime temperatures. Additionally, power line (transportation and distribution) losses ($I^2 R$ losses) are much lower at night due to lower amperage levels in the conductors and transformers at night than during the day. Taken together, nighttime power is generally less polluting than daytime power; however, this depends on the utility power generation fuel mix.

For example, some utilities use base-loaded nuclear power plants, and, therefore, with respect to stack emissions, nighttime power is very clean (nuclear power plants, of course, have other serious environmental issues beyond stack emissions). However, some nighttime power is actually dirtier than daytime power; for example, with utilities whose nighttime base electric load is supplied by coal-fired power plants but who use natural gas fired power plants to meet customer peak demand during the day. Studies have shown that for every kilowatthour of energy shifted to off-peak from on-peak, there is a reduction of 8 to 30 percent in fuel consumption, air-pollutant emissions, and greenhouse-gas emissions produced by the power plant.[1]

Finally, as less cooling equipment capacity, which often use refrigerants, is needed with thermal storage systems, less refrigerant leakage occurs, which may be harmful for the environment (destruction of the ozone layer, for example) depending on the refrigerant used.

Benefits to the Utility

During peak periods of the year, it can be an ongoing challenge for some utilities to meet customer peak electrical demand when they do not have enough of their own power generation resources and must buy on-peak electric power from other sources, which is often relatively very expensive. In addition, some cities have highly loaded or constrained electrical grid infrastructures (in addition to the constrained power transmission lines noted previously) that are at or near capacity during peak demand periods. The capacity shortfall challenge is aggravated at high ambient air temperatures, which reduce the transmission capacity of the electrical distribution system. Thermal storage offers a way to shift peak cooling production, which can represent as much as 40 percent of a facility's electric demand,[2] from the constrained on-peak periods to off-peak hours, when excess utility capacity exists in both power generation resources and in the infrastructure grid (the wires could handle more electrons). The use of thermal storage can delay the need for additional utility power plants and additional transmission network capacity. Utilities may even be willing to negotiate better electric rates due to the thermal storage facility's improved load factor.

Society Benefits

Society benefits from the use of thermal storage for some of the reasons listed above, and the current (2010) U.S. Congress is presently considering incentivizing the installation of thermal storage. Society benefits from less pollution and CO_2 emissions from power plants due to lower facility energy use, less transmission loss, and possible more efficient power plant operations for nighttime produced power. Modern society uses tremendous amounts of electricity and all projections show a continued increase in peak electricity demand. Utilities and governments are looking at investing billions of dollars into upgrading and expanding existing electrical infrastructure grids, creating smart grids, and developing additional power generation resources, increasingly including those from alternative energy sources such as wind, ocean, geothermal, biofuels, and solar power. Thermal storage can help reduce peak period electric demands, and, therefore, reduce the strain on utility generation resources, electric transmission lines, and city electric infrastructure grids. In addition, as wind power (which is by nature intermittent and often has its greatest output during off-peak nighttime periods) is employed in an increasing share of the electric generation mix, energy storage (thermal or otherwise) must play a correspondingly increasing role in the electric grid. Thermal storage is perhaps the most technologically ready and most economical means of energy storage to contribute to this role.

Thermal Storage Basics

The term "thermal," as discussed in this book, means relating to or caused by heat. One definition of heat is the flow of energy due to a temperature difference; however, heat can also be transferred by radiation (as from the sun) and by direct mixing of fluids (known as convection). From a study of physics, as well as one's own experience, one learns that it takes energy to raise a material's temperature or to cause any material to change phase from a solid to a liquid or from a liquid to a gas. Conversely, energy is released or removed when a material or substance's temperature is lowered or when a substance changes phase from a gas to a liquid or from a liquid to a solid. A given amount of a material with a higher temperature contains more energy than the same amount of material at a lower temperature. Likewise, a given amount of material at a higher state of matter (e.g., as a gas versus as a liquid, or as a liquid versus as a solid) contains more energy than the same amount of material at a lower state of matter.

Therefore, thermal energy can be stored (charged) or removed (discharged) in a material by raising or lowering a material or substance's temperature or by causing a material or substance to change phase. When energy is stored and discharged by a difference in temperature of a material or substance, the process is known as "sensible" thermal storage, and when energy is stored and discharged due to a change in phase (e.g., liquid to ice and ice to liquid, respectively), it is known as "latent" thermal storage. When a material or substance changes phase, usually, there is no change in a material's temperature.

The amount of energy per unit of mass that must be removed from liquid to cause it to turn into a solid is called the latent heat of fusion, and, of course, is the same amount of energy that must be added to a solid to cause a solid to turn into a liquid. In English units, the latent heat of fusion for water is approximately 144 British thermal units (Btu) per pound mass, and occurs at 32°F at 1 atmosphere of pressure.

A common thermal energy storage term is tons or ton-hours. This term is derived from the fact that the earliest space cooling was achieved by melting ice. A ton of refrigeration is the amount of cooling that can be achieved by melting 2000 lb (1 ton) of ice in 1 day. As a rate of cooling, the "ton" unit of cooling capacity is defined by the rate of melting a ton of ice over 24 hours. With the heat of fusion of 144 Btu per pound (lb) of ice and 2000 lb of ice, this equates to 288,000 Btu over 24 hours, which is equal to an hourly rate of 12,000 Btu/h. Thermal storage is measured in ton-hours and 1 ton-hour is equal to 12,000 Btu.

Thermal Storage Media

There are various materials or substances that have been successfully used in thermal storage systems, with sensible water and latent water or ice being the most common thermal storage media. Successful thermal storage media will likely have many of the following material properties:

- Common
- Low cost
- Environmentally benign
- Nonflammable and nonexplosive
- Nontoxic
- Noncorrosive
- Inert or stable
- Easy to contain
- Compatible with HVAC systems
- Transportable
- Easy to get energy into and out of the storage material
- Have properties that work well within the range of temperatures desired for thermal storage
- High thermal conductivity for heat transfer
- High thermal density
- High specific heat (for sensible systems to transfer more heat per unit volume)
- High latent heat of fusion (for latent systems to transfer more heat per unit volume)

Of course, the more common a material, the more readily available it is. And it is likely that the material will be of relatively low cost. Ideally, thermal storage media should not harm the environment if it were to leak to the surroundings. Nor would one want thermal storage media to be dangerous or toxic to plant operators, or to be corrosive to plant systems. The ideal thermal storage media should be relatively inert and stable over time and require little or no attention. A relatively high thermal density (mass density times specific heat) is a positive attribute for thermal storage media because it allows relatively more heat transfer per unit volume, all else being equal. Likewise, a high specific heat is a positive attribute for sensible thermal storage media, because it also allows relatively more heat transfer per unit volume, all else being equal. For latent systems, such as ice thermal storage, it is a positive attribute for thermal storage media to have a high latent heat of fusion because it allows more heat to be stored per unit volume. It is apparent from a review of the ideal thermal storage media properties outlined above that water exhibits all of these properties and that is why water (or water-ice) is the most common thermal storage media in use today.

The following are typical materials used for thermal storage media and thermal storage systems:

- Water
- Ice or ice crystals
- Low-temperature (usually aqueous) fluids (LTF)
- Phase change materials (PCM) such as paraffin waxes
- Molten salt
- Gravel
- Soil
- Bricks
- Concrete
- Building mass or structure itself

Ice (the solid form of water) is one of the more common thermal storage materials used today. Ice provides one of the highest theoretical thermal storage densities (and, therefore, the lowest storage volumes) of any thermal storage system and uses one of the least expensive storage materials, water. Ice systems have been shown to be excellent for smaller and even for some larger packaged installations. The primary disadvantages in ice-based thermal storage systems involve the often special ice-making equipment required for the systems, the often greater energy consumption inherent in operating the lower temperature equipment, and a typically shorter equipment life than competing sensible chilled water solutions.

PCM include the true PCM such as paraffin waxes, organic waxes, and salt hydrates (water or ice is, of course, a PCM but is typically considered as a separate category). PCM also include materials that combine to change phase at a given temperature point, such as eutectic solutions, and materials that form high energy bonding structures. Research is presently under way to develop advanced compact phase change thermal storage materials and to investigate thermo-chemical materials and methods (e.g., sorption).

Thermal storage can be obtained in the temperature changes of almost any material. Many thermal storage systems have been built using gravel. Gravel has the advantage of being able to produce air temperatures very close to the storage temperature because of the large surface area of thermal contact. Gravel thermal storage systems in some cases can be relatively inexpensive. Systems using gravel use cooler night air or evaporative cooling for production of cooling. In some cases, air quality problems may result, including damp and musty odors. In certain cases, tests have shown that the gravel systems require more energy than conventional cooling solutions.

Concrete or other building materials are sometimes used to provide at least part of the building cooling during on-peak energy periods by sub-cooling a building during the night or outside the peak electric demand period. The idea is as old as buildings themselves. In areas with substantial daily temperature swings, the windows are often opened. In other cases, buildings are sub-cooled to reduce the load at a peak period. When buildings are conditioned around the clock, the only variable is the outdoor heat gain factors. The mass of the buildings averages the effect of changes in outside air temperature and solar gains.

The earth or rock strata underground are sometimes used for thermal storage. Most of the work in this area is aimed at long-term or seasonal thermal storage. For example, a supply well and a recharge well combined with a closed-loop cooling tower might cool water during winter and, over a long period, reduce the temperature of a confined underground gravel strata or contained aquifer. During summer, circulation of water through the strata could exchange energy to provide chilled water. There are many variables with this system and much research is needed to develop optimum applications. Several projects using seasonal storage in aquifers have been designed in the U.S. Pacific Northwest, Canada, Sweden, and the People's Republic of China. One of the oldest in operation was installed in Portland, Oregon, in 1947.

Types of Thermal Storage Systems

As discussed in Chap. 3, and in other chapters of this book, there are many different types of sensible and latent thermal storage systems including the following:

- Chilled water systems—stratified chilled water thermal storage and multiple tank arrangements
- Ice storage systems—internal melt and external melt
- Hot water thermal storage systems
- Aquifer thermal energy storage (ATES)
- Underground thermal storage, which includes ATES, underground storage tanks, pits, bore-hole thermal energy storage, and man-made underground wells and structures
- Brick or ceramic electric storage heaters
- Under-floor thermal storage systems
- Solar thermal storage

Additionally, there are many different types of arrangements, and thermal storage has even been achieved by collecting and piling up snow in winter and covering it up with reflective tarps to be used for cooling throughout summer.

Chilled water thermal storage systems are sensible thermal storage systems, and thermal storage is, therefore, achieved by a change, or the difference, in the chilled water temperature, that is, the temperature difference between the "warm" chilled water return temperature and the "cool" chilled water supply temperature. As the volume of chilled water thermal storage to meet a given ton-hour capacity is a function of the temperature difference (delta-T), chilled water systems with a higher delta-T require smaller thermal storage volumes and are subsequently more economical than thermal storage systems with relatively lower chilled water system delta-T. System chilled water delta-Ts are typically in the range of 15 to 30°F; however, some chilled water systems have been designed with lower delta-T (higher delta-Ts are better).

Chapter 5 describes different types of ice thermal storage systems including the following:

- Ice harvesters
- Ice-on-coil systems (internal melt and external melt)
- Encapsulated ice systems
- Ice slurries

An ice harvester is similar to the icemaker seen in motels and restaurants and is one of the oldest types of active thermal storage systems. The most common type of ice harvester systems produce ice on refrigerated plates or tubes, which, through the process of periodic defrosting, is allowed to fall off the plates or tubes and drop into a water-ice mix in a tank below.

An ice-on-coil thermal storage system typically has heat exchange tubes immersed in a water storage tank with refrigerant or water-glycol or brine flowing through the inside of the tubes. Ice forms on the tubes during the charge cycle and is then melted during the discharge cycle. One disadvantage with ice-on-coil systems is that the ice acts as an insulator and reduces heat transfer the thicker the ice becomes. Encapsulated ice uses small dimpled containers filled with water and surrounded by a flowing water-glycol or brine solution; the encapsulated water freezes during the charging or storage process and thaws during the discharge process.

ATES has been used successfully by a number of countries, including Sweden and China who have a large number of successful projects that provide cooling to business and industry and heating to residences. In fact, China demonstrated an ATES system in Shanghai in 1960 when cold water was injected into the ground to stop subsidence, after which it was discovered that the cold water could actually be stored in the aquifer for later use.

Brick storage heaters (or electric storage heaters) use electric resistance heat dispersed in an arrangement of a dense mass of the bricks to store thermal energy. Likewise, gravel can be used for an under-floor (actually under one's house or building) thermal storage system.

Thermal Storage Operating Strategies

As shown in Chap. 7, thermal storage systems are usually sized to be either full storage or partial storage systems. A full storage system provides all the cooling or heating required in a specific period with all the production equipment shut off. This time period is usually the on-peak hours when electrical costs are highest. During nonpeak hours, the production equipment is operated at a higher capacity than that required to meet the instantaneous thermal loads. The surplus heating or cooling is stored in the thermal storage system.

Partial storage systems typically use equipment operating at a constant rate over the entire day (24 hours), when serving peak design day load profiles. For example, facility peak cooling demand can be met with partial storage systems provided from a mix of stored cooling plus chiller-produced instantaneous cooling. During periods of lighter loads, a portion of the chiller output is stored in the thermal storage system. The minimum production capacity is equal to the sum of the loads over a design period such as a day or a week divided by the number of hours in the design period. The production capacity then is the average load for that period.

Full storage systems gain their major advantage from the difference between on-peak and off-peak electric demand charges and energy rates. Transferring demand and energy consumption out of the periods of the day when it is most costly can achieve large energy cost savings, but the amount of cooling equipment and thermal storage capacity needed is generally greater with full storage system than it is with a partial storage system. As noted, a thermal storage system designed to operate as a partial storage system on a peak design day can often operate as a full storage system on days with sufficiently lower loads than peak design.

Thermal Storage Economics

Favorable economics for any endeavor means that a favorable return on investment (ROI) is realized or is projected to be realized based on careful study and planning. What is considered a favorable return, of course, varies from person to person, from company to company, and varies over the short and long term depending upon current economic conditions. The bottom line is that capital (i.e., money) is or will be invested with a minimum expected financial payback (a certain amount of money returned over a certain amount of time). However, in some cases, thermal storage systems represent both the lowest first cost alternative as well as the alternative with the lowest energy costs (or greatest energy cost savings).

As discussed in Chap. 8, there are a number of key economic factors that affect any financial analysis such as the following: the discount rate; the length of the time period used in the analysis; and estimated escalation rates for utilities (electricity, natural gas, other fuels, and water), labor, and maintenance. Further, thermal storage economics depend on a number of important factors including both technical and economic factors. Initial capital required for construction (known as first costs) directly impacts the bottom line of any life-cycle cost analysis and is a major factor whether or not a proposed thermal storage system is more economical than proposed alternatives. The construction costs, of course, depend upon a myriad of factors too numerous to enumerate here, and there is a wide range of construction costs even normalized to a per ton-hour basis. The cost required to install a piping distribution system can be significant, especially if there are challenges such as numerous existing underground utilities (known and unknown).

Key technical factors that affect thermal storage economics include the following: the size of the delta-T for sensible thermal storage systems (the larger the delta-T, the more economical the system—this is true for all hydronic systems, not just thermal storage systems); the availability of space for a thermal storage system; the type of space available for a thermal storage system (e.g., open land above ground, open land below ground, space in mechanical rooms, space in or under parking garages); the location of the thermal storage system relative to the central plant; the type of thermal storage system installed; and the thermal storage operating strategy (full or partial shift) employed. Full storage systems provide the maximum operating cost savings; however, partial shift systems utilize smaller thermal storage and smaller equipment, and thus often provide the lowest capital cost, coupled with some operating cost savings, often yielding the best life-cycle economics. However, actual conditions such as hourly load profiles and utility rate structures must be analyzed to determine the optimum design solution for any given application.

History

Thermal storage occurs whenever a heating or cooling process is performed at one time while energy is stored to accomplish heating or cooling at a different time. Of course, some thermal storage occurs naturally, and some of the earliest forms of thermal storage were designs that built on those natural processes. As an example, people found that massive stone structures created a thermal flywheel effect. A stone fortress, with its thick walls, did not get as hot in the day as the outside, or as cold at night, even without any artificial heat or cooling added. This fact led to buildings that purposely used mass to accomplish passive thermal storage.

Some 4000 years ago in Iraq, it is reported that communities used natural ice for cooling,[3] and even today in northern Iraq, Kurdish communities still continue to use natural ice, which is stored in earth pits for use during summer. Later, annual thermal storage was achieved when the Roman emperors and wealthy Romans sent slaves into the Alps to gather ice blocks from glaciers and bring them to Rome for use during summer for cooling buildings and cooling some desserts. Victorian era houses used blocks of ice cut from frozen lakes during winter for cooling during hot summer months. The process of using naturally produced ice was expanded, especially in the United States in the later 1800s using ice from the Great Lakes region. During winter when the lakes froze over, men took ice saws and cut blocks of ice, which were then stacked in large piles covered with sawdust to prevent melting. This pile of ice was thermal storage because it was stored for later use. The ice was removed from the pile during summer and delivered to customers. Some ice was used for space cooling, but it was primarily used for food preservation keeping food cool in specially constructed ice boxes, beverage chilling, and hospital use.

When mechanical freezing processes were developed on a commercial scale in the early 20th century, which largely put the natural ice harvesting companies out of business, refrigeration systems were typically limited to large central plants and large ice plants that produced block ice, which was delivered to homes even up to the 1940s in the United States.

Another very early form of thermal storage was hot water storage. Some of the earliest forms of thermal storage were large tanks that were heated from a fire located under them. With the development of more sophisticated plumbing systems, hot water systems were designed with storage tanks. Hot water storage tanks were included for one or more of several reasons. Usually, the reason to add hot water storage tanks was either to provided large quantities of hot water for a short duration using smaller heating systems than would be needed for instantaneous heat, or, in the case of homes, to use heat from the kitchen stove to heat and store hot water when the stove was being used. The stored hot water could then be used later when the stove was not in use. Most of the early systems took advantage of the fact that hot water compared to cold water is less dense and, therefore, lighter than the same volume of cold water. With a properly designed piping system, the difference in density between the water at different temperatures would cause a natural flow of hot water from the stove back to the top of the tank, with the denser colder water in the bottom of the storage tank flowing back to the stove. Hot water storage tanks are almost universally used today in residential buildings. In some cases, electric hot water tanks are controlled by the serving utility to hold off the electric heating element during peak electrical demands such as morning warm-up. During that period, hot water was provided from stored hot water, which had been heated previously.

In 1940, MIT constructed a house with solar heating that had a thermal storage tank in the basement, which was able to keep the occupants comfortable even in winter.

Gravel has been used as a storage medium for both heating and cooling storage. One example is a state office building for California in Sacramento. Another is a church in Springfield, Oregon, and a solar home near Cottage Grove, Oregon. In 1959, a system was installed in Denver, Colorado, that used glass plate solar collectors and a pea gravel bed for thermal energy storage.

Additionally, a few buildings have been designed using the building structure itself as a mass storage. One such building is an Oregon State University science building constructed in the late 1970s. The floors are constructed with hollow cores and cold outside air is drawn through them at night for cooling storage or exhaust air discharged through them for heating storage. Several such buildings have been designed in Europe.

Some early air-cooling systems used cold river or lake water pumped through coils or sprayed into the air stream to provide space cooling. With the invention of refrigeration systems, mechanically generated chilled water replaced cold river or lake water for the same use. One of the first notable uses of mechanically cooled air was in movie theaters. Movie theaters would often announce the presences of air-conditioned space even more predominately than the movies they were showing. The large difference in cooling load between the hours when a theater was in use and the other hours when a theater was not in use led to an operational technique to reduce the size of the cooling equipment needed. When not in use, the space was precooled colder than the desired comfort temperature. This meant some discomfort when people first arrived and, by the end of the day's use, the space temperatures were higher than comfortable. This process of using the building's mass for thermal storage is essentially a flywheel cooling process. However, the size of cooling equipment needed is smaller than that needed for the peak instantaneous loads. A similar technique is used today in some large sports arenas.

By the 1920s, Willis Carrier had developed a centrifugal chiller to chill water for distribution. The first centrifugal chiller is on display in the Smithsonian Museum in Washington, D.C. Chiller equipment proved to be reliable long-life compression refrigeration-cycle machines. A chiller machine with a serial number in the 300s was still in use in the Los Angeles Times building in the 1980s. Of course, many chiller design advancements and improvements were made in the 60 years that this chiller had been in service.

Early chilled water systems had to meet the cooling load they served at the time the load occurred. For a system serving a high peak load, this meant large costly equipment, heavy electrical loads at peak times, and much lower or no load at other times.

Some of the earliest space cooling thermal storage systems ever designed were for churches. A church might have a relatively huge cooling load Sunday morning for a couple of hours and little or no cooling load for the rest of the week. By the 1930s, a few church systems were designed using chillers operating all week long producing and storing chilled water for use during the peak hours Sunday morning. This thermal storage solution only required a chiller perhaps 1/80 of the size of a chiller sized to meet the peak load when it occurred. However, the thermal storage process also needed large volumes of chilled water since, at that time, nearly all chilled water cooling systems used 8 to 10°F temperature differences (delta-T). This meant that very large chilled water storage volumes were needed, and also the small temperature difference meant that the resultant small difference in density made it very difficult to circulate water in a thermal storage tank, without a mixing of the colder supply water and the returning warmer chilled water, which degrades the thermal storage capacity.

The fairly poor performance of early chilled water thermal storage systems and the high cost of large storage volumes needed discouraged much development of chilled water thermal storage systems for buildings until some factors changed later on. However, thermal storage continued to be used in dairy farms, churches, and movie theaters, all applications with traditionally short-duration high peak loads.

One factor that encouraged the development of thermal storage for buildings was related to building heating needs. In the U.S. Pacific Northwest, Bonneville dam was built in the 1930s, and that was followed by a series of other hydroelectric dams on the Columbia River. Those dams provided large amounts of relatively low-cost electrical energy. The Northwest tends to be cool most of the year, and the typical heating and cooling system for large buildings use central air-handing units. The large central systems typically mixed return-air with colder outside-air to a usual blend of 55°F supply air. That mixture was cool enough to provide the cooling needed by any space without mechanical cooling. Spaces that required less cooling or required heating had heat added. The concept was called "free outside-air cooling," since one did not have to pay to operate cooling equipment, but perhaps a better name was free heat rejection. The center of any large building, without exposure such as a roof, ground floor, or outside wall can only have heat gains, such as that from people, lights, and equipment. On the other hand, spaces with external exposures can require either heating or cooling depending on outside conditions. It is typical for older large buildings in cool climates to require heating on the perimeter most of the year, and, at the same time, have a heat surplus in the central building core. This fact led to a widespread application of the so-called "bootstrap heat pump." J. Donald Kroker was one of the leaders of its use, dating back to the years following World War II, and his designs were based on systems he had seen in Europe while he served in the military. A bootstrap heat pump eliminated the so-called free cooling cycle and instead provided mechanical cooling with chilled water to the spaces that required cooling such as in the interior core. Outside air was kept to the minimum needed for ventilation. The chilled water was produced by a central centrifugal chiller with a double-bundle water-cooled condenser. A water chiller uses the refrigeration cycle to take heat from the chilled water (and thus cools the chilled water) and raises the refrigerant in temperature to a point high enough to be able to reject the heat to the ambient, such as with condenser water in a cooling tower. A typical standard chiller might have chilled water from 55 to 45°F and rejected the condenser heat to cooling tower water at 95°F. Water from the cooling tower is cooled by evaporation and returned to the chiller at perhaps 85°F. The typical bootstrap heat pump would provide greater temperature lift with perhaps 120°F water temperature leaving the condenser and perhaps 100°F entering water temperature. This warmer condenser water was used in specially designed coils to provide the building heat. The heat provided to the building included the heat removed from the building core and the additional equivalent heat energy work output from operating the chiller.

This higher condenser temperature resulted in higher energy consumption per ton removed than in a typical cooling process. A typical process might require up to 2 percent more energy per °F higher temperature of the condenser water. Still, even with that penalty, the process from a heating perspective is very efficient with a coefficient of performance (COP) of 3 or better. That means 1 kWh of electric energy provides heating equivalent to 3 kWh of electric heating. Moreover, that combined with very low-cost electric power resulted in a very cost-effective combined heating and cooling solution. As a consequence, many larger buildings had those heat pump systems installed.

One drawback of the bootstrap heat pump solution, however, is the fact that heating and cooling cycles for the buildings do not typically coincide. In the early morning, buildings tend to have high demands to heat up the structure and meet other heat loss, and, at that time, have little if any cooling demands on the interior. Later in the day, with warmer outside temperatures, solar gains, lights, people, and processes, there tends to be more heat available than needed by the building. The issue of dealing with noncoincident heating and cooling loads when using a heat pump solution led to the use of thermal storage by the 1970s.

The systems simply added chilled water thermal storage to the same heat pump systems just described. The heat pumps then were used primarily as needed to meet the heating load drawing heat from the chilled water in the storage tank. The stored chilled water was then used later in the day to meet surplus cooling loads and in the process store heat in the chilled water tank to meet the next day's heating loads. A serendipitous effect was the fact that the chiller did not have to be as large as the peak cooling load. In fact, the design typically spread the cooling load evenly over a 24-hour period. The reduced cooling plant size often resulted in a system that cost less than a conventional nonstorage system even with the additional thermal storage tank cost. Lower capital costs occurred because of much lower chilled water equipment cost and the lower related building, piping, cooling tower, electrical service, costs, and so on.

In the latter 1960s, Mt. Hood Community College was built around a central campus heating and cooling process using that approach and included thermal energy storage. A number of other larger schools and office buildings were designed with such heat pump or thermal storage systems. Because the volume of stored water is large, pressure tanks would be prohibitively expensive. The thermal storage tanks systems, therefore, tend to be atmospheric pressure type tanks (i.e., nonpressurized).

In the mid-1970s, a somewhat different approach was designed for Weyerhaeuser near Tacoma, Washington. A pressurized tank of warm condenser water was designed into the system. During the afternoon surplus gain period, some of the condenser water was stored in the tank. If more heat was needed by the building, an electric heater could heat the water even more with low-cost off-peak electric energy. During startup and high heating periods, the heat from the stored water was introduced into the chilled water system to increase the cooling load and thus the heat was available from the system for building heating. The primary differences were the thermal storage tank was smaller, but more costly being a pressure tank, and the chiller was larger because it had to meet the peak cooling load.

During the 1980s, similar solutions were designed in Wisconsin using ice as a storage medium. The ice was used for peak cooling, and heat was produced and utilized at night as ice was produced.

Many of the systems during the 1960s to 1980s used large concrete underground structures that resembled basements but were filled with chilled water. Those systems tended to have poor separation between warmer and colder chilled water. This was due, in part, to the fact that stratified systems tend to have a 2- to 3-ft zone of mixed water separating the colder (and denser) chilled water supply at the bottom of the TES tank from the warmer (and less dense) chilled water return on top of the thermal storage tank. The underground thermal storage tank design tended to be 15- to 25-feet (ft) deep, and 2 to 3 ft is a significant percentage of the volume if the tank is only 20 ft deep. Some smaller systems were designed using fiberglass storage tanks similar to gasoline underground storage tanks. Most such systems used several tanks in series to provide separation between the chilled water supply and chilled water return.

During the 1920s, a number of large buildings in Toronto, Ontario, Canada were designed with large underground concrete storage substructures. In an effort to improve stratification, a patented system was developed to introduce a membrane diaphragm between the chilled water supply and return within the TES tank.

A different approach was used in the Lane County, Oregon, Courthouse in 1978. There the tank stratification was improved by using a tall vertical steel storage tank. The tank was 65 ft tall, and as a consequence, the 2 to 3 ft stratification zone was a much smaller part of the total tank volume. The thermal storage tank performance was also enhanced by providing a greater temperature difference in the chilled water system than the typical 10 to 14°F used commonly at that time. Using a 40°F chilled water supply temperature and a 64°F chilled water return temperature (and therefore a 24°F delta-T), the thermal storage system only required half the volume of chilled water that would have been needed with a then standard 12°F delta-T system.

By the mid 1980s, thermal storage systems were being designed as purely chilled water storage without any use of the heat rejected by the chillers. The driving force for this change from previous designs was largely cost incentives offered by utilities for systems that would transfer the maximum amount of electric load from on-peak periods to off-peak periods. Many of the thermal storage systems designed were full storage systems. As previously noted, that means that the chillers were held off entirely during a certain on-peak period of highest demand, and cooling was provided entirely from chilled water previously produced and stored during the off-peak hours. To encourage design of such systems, utilities provided capital toward the first costs (generally proportional to the on-peak electric load demand reduction), and developed special electric rate structures. Those incentives often changed the economic advantage from partial storage solutions to full storage solutions. Many full storage systems of that type have been designed and have proven to be cost-effective especially on large central systems such as those with colleges, universities, or district cooling systems.

A different type of utility incentive for Lane Community College resulted in the campus being designed to be heated with two 3000-kW electric boilers in the mid-1960s. At that time, the local electric utility offered a rate of $0.003/kWh, with the agreement that the utility could require one boiler to be turned off at the time of peak utility demand. For the utility at that time, their peak electric demand was largely heating driven, with a peak load from 6 to 8 AM. As the electric load increased over time, the utility modified that agreement in a way that encouraged other solutions. A bootstrap heat pump was installed along with a large heating thermal storage system. The heating thermal storage system used 200,000-gal underground pressurized stainless steel insulated storage tanks. Water in the thermal storage tanks was heated to 240°F by the boilers at night and used to provide additional warm up heat the next morning. The tanks were obtained from the government as surplus from an early missile development program.

From the late 1980s forward, stratified chilled water storage has been the dominant form of sensible cool storage, supplanting the earlier forms, which included the multiple tank (or "empty tank") designs, the diaphragm or membrane tanks, and labyrinth tanks. Additionally, from the late 1980s forward, most chilled water tanks are specified and procured as "turnkey" installations complete with the internal flow diffuser systems for proper thermal stratification and inclusive of guaranteed thermal performance, versus earlier installations with separately designed and installed diffusers.

Stratified chilled water storage systems in the past have been limited to chilled water applications of 39°F or greater chilled water supply temperature, and typically have had about a 20° temperature rise (delta-T) at best. A couple of developments have increased the range of temperatures. 39°F has been the limit on the low side due to the fact that 39.4°F is the maximum density point for pure water (pure water becomes less dense at temperatures below 39.4°F). A patented solution is available to add chemicals that lower both the temperature of maximum density and the freezing point below that of pure water. Low-temperature fluid (LTF) thermal storage systems with 30° or lower chilled fluid supply temperature have been installed and are discussed in Chaps. 4 and 9.

By the 1970s, some systems were being designed using ice and other PCM. Early ice storage systems (ice harvesters) tended to be designed around some method of freezing ice on plates or on tubes with direct expansion refrigerant on one side of the plates or tubes and water freezing on the other side. As the water freezes, it adheres to the plate or tube and forms an insulating layer, which slows down subsequent heat transfer. Most systems had some way of removing ice from the plates or tubes such as defrosting. Defrosting tended to introduce more energy inefficiency into the process. Some alternative ice thermal storage systems had large piping systems with pipes in a water bath and refrigerant in the pipes. Such systems tend to require large volumes of refrigerant and lose ice production capacity with the increase of ice coating on the pipes. A trend today is that most ice systems use chillers and a water-antifreeze solution to freeze water inside capsules or on the outside of pipes in a water bath storage vessel.

Early thermal storage work also developed several eutectic storage solutions. These were used on both heating and cooling storage systems. Eutectic storage systems are latent storage systems and as with ice tend to provide most of their thermal storage within a single temperature point or at a very narrow range of temperatures. Most typically, the eutectic solution was in some capsules similar to encapsulated ice. The major difference was that the circulating solution used was water, and standard chillers were used producing chilled water at about the same temperature and cost as conventional nonthermal storage systems. There have been many problems with those solutions, including failure to function properly over time and high circulation volumes in comparison with stratified chilled water solutions.

There also have been several systems that drew on natural processes to provide longer term storage, typically annual. One such solution drew freezing outside air in winter into a building size insulated structure, and water was sprayed into the air producing ice crystals inside the structure. The ice crystals were stored in the building for cooling during summer. Similar systems have circulated water between wells, and the water cooled using cold winter air, which produced lower well water temperatures to use in summer for cooling.

A fairly recent technology development generates ice slurry in a vacuum vessel at the triple-point of water (where solid, liquid, and vapor phases are all in equilibrium). The refrigerant is the water vapor; and as there is no heat exchanger surface between the refrigerant and the water-ice, it has low energy consumption compared to conventional ice-making systems. The technology has been applied for thermal storage, as well as in district energy systems (as a heat pump), deep mine cooling, and warm weather snow-making applications.

Some systems today have naturally produced annual storage. Cold water generated by winter temperatures in the Great Lakes or other lakes is taken off the bottom of the lake in summer and pumped through plate-and-frame heat exchangers and then returned to the top of the lake. This lake cooling can produce chilled water for very low costs. Similar solutions are under design in some very hot regions such as Hawaii and parts of China, where, despite the hot climate, brine water temperatures below 40°F are available at great ocean depths near the shore. Although these systems appear to be very cost-effective over time, they may not be properly thought of as thermal storage in the sense that man did not construct the process that provides the cooling or the thermal storage.

Recent developments include concentrated and parabolic trough solar power systems, which use molten salt, hot oil, or hot rocks in their process that can be used as a high-temperature thermal storage system for use in the evenings and overcast periods.

Key Issues Facing Industry Today

Some of the key issues facing the thermal storage industry today, as challenges or opportunities include the following: the ever-increasing demand for electricity; the need for energy storage in conjunction with alternative energy technology including especially wind energy; the need to better understand utility emissions at night versus utility emissions during the day depending on one's particular location and one's utility company; the debate about utility incentives as to whether incentives help or hurt the thermal storage industry; proposed thermal storage legislation in the United States; and educating the public and policymakers regarding the many important benefits of thermal storage to facility or building owners, utility companies, and to people and their nations.

Total electrical energy consumption has trended up throughout most of the world for more than half-century a century. In the United States, per capita electrical energy use has increased from approximately 4000 kWh per year in 1960 to approximately 14,000 kWh per year today.[4] The on-peak electric load growth is increasing faster than the off-peak load growth; and much of the peak growth (past and future) is related to increasing air-conditioning or cooling loads—loads that can be addressed via cool thermal storage. It seems as if every day there is a new device that needs to be plugged into a power outlet. With an increasing population and a tripling of the per capita electrical energy use over the past 50 years, increased electric power demands beyond those envisioned have caused challenges for utility companies as they work to meet customer demand, with sometimes insufficient generation resources and highly loaded infrastructure grids that are at or near the limit of their current carrying capacity. As discussed in this chapter, thermal storage benefits utility companies by helping to reduce on-peak demand growth by shifting production to off-peak periods, thereby deferring or delaying costly electric infrastructure improvements and investments.

Thermal storage has an important role to play in the success of alternative energy sources. For example, wind and some solar energy are often not constant or predictable. In fact, peak wind turbine output does not coincide with peak electrical demand; as one can imagine, a utility experiences peak electrical demand due to building air-conditioning on a hot, windless day. Cool thermal storage can allow for air-conditioning production at night, when wind power is often in excess of demand, for use during the day when wind power availability may be less. One might imagine a situation where the local utility could have the ability (via the "smart grid") to dispatch (discharge) regional thermal storage systems from individual buildings and facilities as needed to reduce peak power demand; thermal storage that was produced the night before using inexpensive nighttime wind power to generate cooling. Additionally, as discussed in this chapter and further in Chaps. 3 and 6, hot water thermal storage is an integral part of many solar thermal systems, as it is with most domestic hot water systems. Utility companies and the thermal storage industry sometimes travel in different circles, and one mission is to build more bridges between the respective organizations.

As noted in this chapter, calculating the expected emissions credit from the use of thermal storage is presently not straightforward, as the amount of utility emissions at night versus the day depends upon location, utility company provider, and the mix of power generation resources (e.g., fuel oil, natural gas, hydropower, nuclear power). For example, a utility that uses nuclear power plants to meet their base electric load will obviously have much less stack emissions at night than a utility company that obtains its base electric load from coal-fired power plants. Presently, only average emissions data are available for a particular utility company or utility region but not how those utility emissions change from day to night or from season to season. Those in the thermal storage industry need utility company hourly emissions data so that facility or building owners, planners, sustainability professionals, and thermal storage engineers can quantify for stakeholders the environmental benefits of installing thermal storage.

There seems to have been a debate over the many years as to whether utility financial incentives help or hurt the thermal storage industry. Basic economics seems to indicate that if more dollars are supplied (e.g., by providing utility financial incentives) then, all else being equal, demand should increase; utility incentives have helped fund many thermal storage projects, some of which would definitely not have proceeded without the utility incentive. However, the skeptical nature of human beings, being what it is, leads people to sometimes say or think that if utility needs to provide an incentive, then thermal storage cannot stand on its own, which, unfortunately, has been a barrier to the installation of thermal storage. Thermal storage provides financial benefits to utility companies, and, therefore, they also have an incentive to support the installation of thermal storage. Nevertheless, a great many thermal storage installations have proven economically very attractive even in the absence of utility incentives, though generally these are partial storage systems and are typically built only at times of new construction, major load growth, or chiller plant rehabilitations. A much more widespread use of full shift storage systems, including those constructed as pure retrofits (i.e., not associated with cooling load growth or chiller retirements), can be foreseen only with utility system investment or cost sharing. Continuous education of various stakeholders is always needed.

The Storage Technology of Renewable Green Energy Act of 2009—in the year 2009, legislation was introduced in the U.S. House of Representatives that would provide a 30-percent tax credit to individuals and business when they install thermal storage systems. As discussed, thermal storage provides benefits to the public, including less pollution and CO_2 emissions, and less strain on our utilities and electrical infrastructures in the face of ever-increasing electric demand. Companion legislation has also been introduced into the U.S. Senate. Unfortunately, progress on the bill has been limited due to the present (2011) political climate in Washington, D.C.

As already discussed in this chapter, implementing thermal storage has many benefits for multiple stakeholders and often helps to reduce facility energy and to reduce power plant fuel consumption, pollutant emissions, and CO_2 emissions. The use of thermal storage allows for smaller refrigeration equipment, and, therefore, has lower refrigerant leakage rates, which may damage the atmosphere. Therefore, thermal storage is a "green" energy solution that also saves facility or building owners energy costs over time. Unfortunately, at the present time, most of the public does not even know what thermal storage is, let alone the benefits of thermal storage as a green technology. It is necessary for those in the thermal energy storage industry to educate and advocate for the installation of thermal storage so that its many benefits are understood by and distributed to all those for whom the benefits of this proven technology can and do accrue.

References

1. Tabors Caramanis & Associates, *Source Energy and Environmental Impacts of Thermal Energy Storage*, California Energy Commission, 1995.
2. Bush, R., Utilities Load Shift with Thermal Storage, Transmission & Distribution World, Energy Storage Supplement, August 2009.
3. "Iraqi Homes Show US How to Build," Aljazeera.net, http://english.aljazeera.net/archive/2005/02/200849154726154132.html (accessed Feb. 04, 2005)
4. Kandel, A., M. Sheridan, and P. McAuliffe, A Comparison of per Capita Electricity Consumption in the United States and California, ACEEE Summer Study on Energy Efficiency in Buildings, August 17–22, 2008, Pacific Grove, California.

CHAPTER 2

Applicability of Thermal Storage Systems

Introduction

As discussed in Chap. 1, installation of thermal storage is most often applicable and economical to facilities and buildings when thermal loads vary over time, where energy and demand costs are time dependent, when new facilities are planned, or when facility expansions or upgrades are needed. Further, thermal storage is applicable to a wide variety of facilities and buildings including commercial, residential, industrial, and institutional facilities. Utility rate structures are a key factor in the application of a thermal storage system and can affect which type of operating strategy, full storage or a partial storage, is more economical. In addition, thermal storage systems are used as part of an emergency backup system, sometimes integrated with fire protection systems, and, importantly, are used with combustion turbine inlet-air cooling (CTIC) systems.

Applicability of Thermal Storage Systems

Thermal storage technology can be (and has been) successfully applied to a wide variety of facilities and buildings as well as occupancy types including, as noted, commercial, residential, industrial, and institutional facilities for both the public and private sectors:

- District energy systems (district heating and district cooling)
- Universities and colleges
- K–12 schools
- Hospitals or medical facilities
- Municipal centers
- Corporate offices or research campuses
- Large commercial buildings
- Data centers
- Military bases
- Jails and prisons

- Pharmaceutical manufacturers
- Telecom, automotive, and aerospace facilities
- Other factories and industries requiring cooling and heating processes
- Residential heating, cooling, and domestic hot water use
- Emergency cooling for mission-critical systems such as data centers and hospitals
- Utility and combined heat and power (CHP) CTIC
- Solar power plants

Each type of facility or building has its own load characteristics and load profile shape, depending on a number of factors (as discussed in the following section), and the size and shape of the load profile directly affect the required size and operation of a thermal storage system as shown in Chap. 7. Different types of facility or buildings and different types of occupancies have very different load profiles.

District energy systems sell or provide heating and cooling (i.e., thermal energy) to downtown office buildings, commercial buildings, universities and colleges, as well as corporate campuses, for example, and, as such, district energy systems often generate power, heating, and cooling in a central plant and then distribute those utilities through a network to individual buildings and users. District energy systems are usually much more efficient than local building systems and require less maintenance than do separate plants in each individual building. Plants that produce electricity and heating for sale are called CHP plants. Most commonly today, CHP plants also produce and sell cooling and are called total energy plants (or trigeneration or combined cooling heat and power—CCHP). Because electrical energy, heating, and cooling are all produced from a single fuel source as part of a combined process, such total energy plants use much less energy than single process plants that use local boilers, chillers, and dedicated electrical production power plants. As an example, waste heat from the plant may be captured and used for heating and cooling production. A CHP plant that is part of a district energy system may use only half of the fuel needed to produce the same amount of heat and power, with the usual practice of buying power from an electric utility and burning fuel in a local boiler. Thermal storage improves central plant operations and economics for a district energy system primarily by (1) reducing equipment size and related infrastructure, thereby reducing related capital cost; (2) reducing ongoing energy costs; and (3) reducing or eliminating on-peak cooling power demand, which reduces net CHP power output just at the time when that power is most valuable. Additionally, heating and cooling are obtained from waste heat generated during the production of electricity, and thermal storage allows the waste energy to be stored as it is available and used when it is needed.

A group of commonly owned buildings, such as universities and colleges, K–12 schools, municipal centers, or corporate campuses, is often an ideal candidate for a district energy system with thermal storage. In fact, today there are an increasing number of colleges and universities that use thermal storage systems (see Fig. 2-1), and many of these facilities are retrofitted with thermal storage to older, existing distribution systems. Where there are existing heating and cooling distribution systems, including thermal storage as a part of the solution, is often the most cost-effective alternative. These facilities with thermal storage routinely save capital costs, defer required equipment installations, and continue to save energy costs year after year.

FIGURE 2-1   Aboveground stratified chilled water thermal storage tank.

Thermal storage is also applicable to individual buildings and has been successfully installed in hospitals, large commercial buildings, and data centers; it can provide for emergency cooling in mission-critical systems such as at data centers and hospitals.

Commercially available products are presently available for residential cooling thermal storage, and residential solar domestic hot water thermal storage is a common practice in many places of the world.

The applicability of thermal storage to CHP CTIC is discussed at the end of this chapter.

Facility Loads

As noted, an important factor in the application of thermal storage is having cyclical facility loads, and the greater the peak load is with respect to the average load over the period of time under consideration (such as 24 hours), the greater the potential is for thermal storage.

In the context of a heating or cooling plant and thermal storage, facility loads refer to cooling loads, heating loads, and electric power loads. For a variety of reasons, the loads or demands on any system typically vary over a 24-hour period and from season to season. For example, most buildings are not used for the full 24-hour day, and often building heating and cooling systems are off during those unoccupied periods. Some parts of a building load, such as lighting gains, occupants, and processes are more or less constant during periods of use. Of course, those same loads will be much reduced or nonexistent during periods of nonuse. Therefore, the facility occupancy, which typically varies over a 24-hour and 7-day period, affects the size and shape of the facility load. Other parts of the facility or building load are determined by factors such as outside air temperature, moisture content of the outside air, and sun angles, which can all continually vary over time. Additionally, other factors that affect a facility or building load are a function of the building itself: shape, construction, color, mass, heat transfer (U) factors, orientation, as well as location (climate).

The cooling load is less than peak conditions when there are reduced or no solar gains (e.g., it is a cloudy day or the sun went down) or when the outside temperature is lower than peak design temperatures. Cooling loads are also less than peak during periods of lower occupancy (less people equals less cooling load). Because of changes in solar intensity, outside air temperature variations, and the more intensive use of the buildings during the day, a typical graph of load versus time (e.g., over a 24-hour period), known as a load profile, shows an increase in cooling load during the day, peaking in the mid-afternoon, and decreasing to a lower level overnight. Cooling and electric loads most often peak in the 2 to 5 PM period. As shown in Fig. 2-2, load profiles typically vary from month to month, and seasonal changes can be large in some climates, with some climates requiring cooling in the summer and heating in the winter.

As noted, additional factors that affect the facility or building load profile include the facility or building: size (floor area); occupancy type or use (e.g., office versus dormitory versus research lab); and operating schedule. A hospital, for example, has a much different occupancy and operating schedule than does a high school. Some

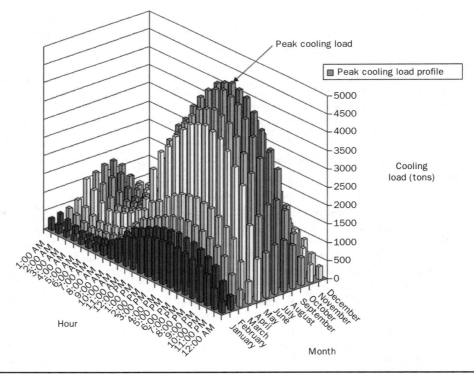

Figure 2-2 Typical average daily cooling load profile by month.

buildings such as office buildings have peak cooling loads that are as much as three times as high as the average cooling load over 24 hours. For such buildings, a thermal storage system can use smaller chillers and store surplus cooling capacity to meet future cooling loads that are in excess of the instantaneous production capability of the installed chillers. A similar case can occur with facility heating loads (however, the shape of the load profile is very different).

Table 2-1 provides data for a different facility's peak-day 24-hour cooling load profile, which as expected varies from a minimum load at night to a peak in the mid-afternoon. As shown, the peak cooling load for the facility is 9092 tons. Also, note that in this case the average cooling load for each hour is summed for the 24-hour peak-day period to provide the peak-day total ton-hours of 123,829 ton-hours. The peak-day total

Peak-day cooling load	
Hour	**Tons of cooling**
1	2,285
2	1,864
3	1,579
4	1,449
5	1,386
6	1,359
7	1,526
8	1,665
9	1,826
10	3,950
11	6,154
12	6,800
13	7,690
14	8,278
15	8,691
16	9,061
17	**9,092**
18	9,018
19	8,910
20	8,551
21	8,622
22	5,935
23	5,401
24	2,735
Total (ton-hours)	123,829

TABLE 2-1 Typical Daily Cyclical Cooling Load

daily ton-hours is an important value for double-checking calculation results and is used in sizing thermal storage systems (larger ton-hours equals larger thermal storage, all else being equal).

The factors affecting cooling load are similar to ones that also affect facility heating loads; however, as shown in Fig. 2-3, heating loads have a load profile that is different in shape from that of cooling loads. In fact, the general rule is that one type of load is the inverse of the other. That is, when building cooling loads are the greatest, heating loads are the least, and vice versa. Most large commercial or institutional buildings, however, usually have some simultaneous heating and cooling loads at all times. In the case of heating demand, there is often a large initial warm-up load when the facility opens, dropping to a minimum load at mid-afternoon, followed by another peak load in the evening before the facility closes for the night. As shown in Fig. 2-3, in this case, heating loads are present only during winter months, and no heating is required for this facility in the summer.

Heating loads can typically include space heating, domestic hot water production, swimming pool heating, as well as process heating. Depending on the occupancy and other factors, as shown in Fig. 2-2, heating loads often peak in the morning and in the early evening with a minimum load during the day. Because of this type of load profile, solar heating with hot water thermal storage works well, as heating can be generated during the day, when solar loads are high and heating loads are low, for later use at night and the following morning.

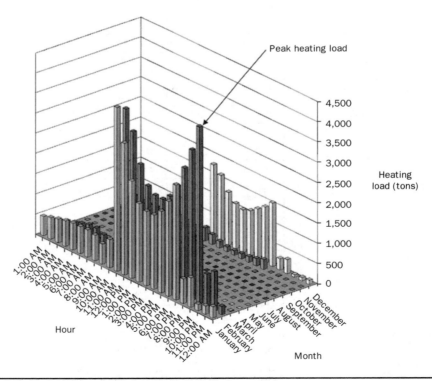

Figure 2-3   Typical average daily heating load profile by month.

Utility Rate Structures

In order to understand the applicability of thermal storage systems to any facility, one must have a basic understanding of utility corporate structures, utility company operations, and different types of utility rate tariffs. A detailed understanding of the applicable utility rate structure to a particular project is required in order to determine projected energy savings. There are several types of electric utilities, and their corporate structure affects to a certain extent what and how utilities charge for power. Utility types include the following:

- Municipal-owned utilities (owned by cities or other units of local government)
- User-owned utilities or co-ops
- Investor-owned utilities
- On-site utilities for a single owner with ownership of production and distribution within that single ownership, such as an industrial plant, university, hospital, or military base

Another difference that affects electric rates is the source of the power sold to customers. Some utilities purchase all the power they distribute from wholesalers. Some call such utilities wire utilities. Other utility companies generate at least a base load in their own power plants. Still other utilities purchase most of their base load but generate power during peak demand using fairly low-cost power plants, which can be quickly started, such as simple CT generators. Many utilities have some combination of sources and both buy and sell power.

Base load power plants tend to be very expensive to construct and require long periods to site, permit, and construct. However, such base load power plants tend to produce low-cost power. Examples are coal-fired, nuclear, and hydroelectric power plants. Power plants that burn fuels such as coal, oil, or gas usually produce superheated steam to drive condensing steam turbines and may be combined-cycle power plants. There are a number of social and environmental challenges to such power plants, and their acceptance is somewhat limited (not in my backyard).

Green energy plants, such as solar, wind, ocean, biomass, and geothermal, are a growing trend both for utility companies and facility owners; however, many technologies may be confined to certain geographic areas and have limitations by their nature as well as social acceptance issues. Presently, most green energy costs more to produce than the base load power plants; however, that may not be true in the not too distant future.

Peak load power plants tend to have lower construction costs, are easier and faster to permit and to construct, but, typically, have higher energy cost per unit of power produced than a base load power plant.

Electrical utilities are considered "natural" monopolies. This is because it is not practical to have multiple power-distribution lines serving the same customer, which effectively eliminates competition for providing electricity to an individual customer. Because the customer can typically buy his power only from the local utility (although this is changing in some locations), states and to a certain extent the federal government regulate the utility companies and control what utility companies can charge for the power they sell. What utility companies can charge is controlled to a certain extent by their physical plant investment (power generation plants, primary power distribution lines, transformers, meters, etc.). Because the government controls what utilities can build, what utilities can charge, and how much profit utility companies can earn, utilities are called regulated utilities. One aspect of this regulation is the development of utility tariff rates, which are schedules of how utilities charge for the power they sell.

The energy rates for residential customers of most electric utilities are fairly simple. Usually, utility companies charge only for the amount of energy consumption and a base cost for service. However, there can also be a base charge for each meter, which is charged regardless of the amount of consumption of electricity for the month. Often, there are blocks of energy consumption that are priced at different rates for each block of energy use. Although usually one pays less for things as one purchases more of the item, in many cases electric utilities actually charge more for each subsequent block of energy usage. Utility customers pay less per unit for the lowest use and consumption block, and above that level, they are charged at a higher rate. The intent is to provide a base level of power to all their customers at affordable rates. This pricing strategy also reflects the fact that the base power load costs the utility less and increased use costs the utility more per unit of power produced.

Many utilities also offer part or all of the monthly consumption as "green energy." This is usually selected by the customer and is presently more costly than other power. Green energy, whatever it is called by the utility, is based on a purchase of power from wholesalers producing power from wind, ocean, hydroelectric, solar, geothermal, or other clean renewable sources.

Some utilities offer special rates for customers willing to allow certain parts of their power to be cut off at peak demand periods. This might mean eliminating hot water heating from 6 to 8 AM, as an example. As noted, in some cases, the serving utility only transmits power bought by the customer from an energy supplier. In that case, the serving utility charges only for services like transmission, transformer, metering, and billing services. The bill passes along the charges from the electric energy marketer.

Industrial and commercial customers usually buy their power under more complex rate structures. These rates are developed by utilities to more closely reflect and control their actual costs in providing power.

A utility has many elements of cost for the power finally delivered to customers. The exact balance of costs depends on the specific utility. In general, costs are either fixed or variable. Fixed costs include transmission lines; transformers; utility-owned generation plants; meters; service and repair shops; basic maintenance and repair personnel; office and management costs; and taxes.

Variable costs include energy consumed in generating power and power purchased from other sources on behalf of the utilities customer. Variable and fixed costs are often melded together in some fashion to provide a certain energy-rate structure. In its simplest form, most commercial and industrial customers pay both for energy consumption and for demand (peak energy usage). Energy, for example, is the number of kilowatthours (kWh) used in the month, and demand is the maximum rate of consumption measured in kilowatts (kW) or megawatts (MW). A simple utility electric rate may not take into account when the maximum consumption occurs. More complex utility electric rates charge more for demand or energy when the utility's demand for power is the greatest, which may result in on-peak and off-peak rates (which are different). Some utilities add a mid- or partial-peak block of hours to each day or month. Usually, these utility electric rates are different for the winter and summer season.

It is important to understand why the rate tariffs are developed and their relationship to thermal storage. Most utilities have a great deal of variation in the rate of power demand throughout the day, day of the week, and season. Time-of-day differences occur between largely residential use during nights and weekends and commercial and industrial use during the daytime of the workweek. Seasonal changes occur, for example, for electric heating during winter and morning periods and for air conditioning use, particularly afternoons of the summer period. A power company has a fixed investment in physical plant; transmission lines, transformers, and generating plant capacity. If the demand for power were uniform each hour of the day, day of the year, and year after year, then all that would be needed is an investment in physical plant equal to that load. However, because, at times, the electric power load exceeds that hypothetical flat load, the investment must be great enough to meet the demand of the maximum served load, and that increased cost to the utility company must be paid in some way. Moreover, as electric peak load increases, the requirement for increased utility company investment to meet the expanded maximum load also grows. This may mean that the utility company must build a new generating plant, increase transmission capacity, or other increased investment in order to provide power, for, perhaps, just a few hours a year. Alternatively, increased peak electric loads may mean the utility ordering more power from a wholesaler that needs to construct or start up a power plant that may only be needed for 15 minutes, for example. This investment or power purchase order may be very expensive.

The option available to the utility is to encourage measures that transfer electrical consumption from peak periods to periods of lower demand, which saves the utility construction dollars or reduces the need to buy extremely expensive power. Utility companies have developed various utility rate structures to encourage this sort of transfer of time of electrical use to shave the peak and fill the valley of electrical demand.

Utility rates that encourage energy use transfer in time are especially attractive to customers with large fluctuations in load or are otherwise able to help themselves and the utility. One such solution is generation of some on-site power at peak times. The other solution is thermal storage. Either solution requires some investment on the part of the facility. The utilities often encourage this investment by passing along some of their savings and penalizing energy use during their peak use period.

For many years, utility companies paid incentives to construct and operate thermal storage at the customer's site, which included cash contributions for capacity that could be generated off peak and held off during the on-peak period. These utility incentives often meant that a thermal storage plant operated nights with cooling production equipment held off in the mid-afternoon, when the customer cooling load was at its maximum, and, which also was the utility's peak demand period. Fewer utilities offer this sort of construction incentive today.

Still nearly every utility encourages this sort of transfer of load in time by the utility rates they offer. Some electric rates penalize energy costs every month of the year by the maximum demand use in any 15 minutes a year. Nearly every utility charges commercial or industrial customers for demand. Moreover, since most customers, with a cooling load, see their maximum demand created by the cooling system adding to other electric loads, thermal storage can reduce demand and thus electrical costs.

The current trend in utility pricing strategies is to shape the electrical costs for major electrical use customers to the actual cost to the utility. The opposite side of that fact is that the customer can be rewarded for reducing energy use when doing so affects the utility in a way to substantially decrease their demand or cost of energy. A term in use today is "real time pricing." With this idea, the power use of the customer is measured and recorded as it occurs. In addition, the actual cost of power generation, purchase, and delivery is measured at that time. The amount of actual cost to the utility is passed on (with markup for fixed costs and profit) to the customer, which can mean that a reasonably small amount of thermal capacity used for even a few minutes a year can achieve great savings.

It is even possible with this rate structure to obtain power with negative costs. "Negative costs" means the utility pays for the customer to accept power. This is possible only in certain situations such as with a nuclear plant that needs to maintain a certain base load or a utility with excess wind power. When the demand by all customers does not equal that base load, it is in the utility or energy wholesaler's interest to pay for energy use at that time.

Electric Rate Schedules

Utility rate structures are often also divided into residential, commercial, industrial, and agricultural customer rate schedules. Utility companies use a number of different types of utility rate structures including (1) tiered rate structures, where different blocks of energy have different costs, with the first block costing the least amount (e.g., the first 500 kWh) and each subsequent block of energy costing more than the previous block of energy; and (2) time-of-use rate structures, where the rates vary by the time of day.

The specific utility rate structure (and hence the utility cost) for a facility often depends upon the size of the service (with larger consumers often obtaining lower utility rates), and, for electrical utilities, the voltage at which a facility or building is served (with customers taking power at higher supply voltages often obtaining lower utility rates). Owning or not owning your own main facility transformer can also affect utility rates, with facilities who own their own transformers often obtaining lower utility rates than facilities where the utility company owns the main facility transformer. Additionally, utility rates may also be affected by a facility's load factor, which is the ratio of average utility consumption rate for a specific period divided by the peak demand for that period, with those with higher load factors, meaning flatter load profiles, sometimes obtaining lower utility rates.

Utility rates can be firm or interruptible, meaning that, if necessary due to insufficient supply available from the utility company, with a short notice (e.g., 15 minutes), the utility company can stop supplying utilities to the facility or building. Interruptible rates are lower than firm rates; however, there is a risk of losing utility supply. The size of the risk depends upon how often a facility's utility company has curtailments (is it frequently or rarely) and upon what the outcome of the utility curtailment means (which could have serious consequences for some facilities).

Special utility rates are sometimes offered for specific groups and uses including to nonprofit organizations, to the elderly, as well as to CHP facilities, since cogeneration is a green technology that can more than double the fuel utilization factor. In addition to normal time-of-use rates, special utility rates are also sometimes provided for thermal storage systems, whereby a utility company encourages thermal storage by offering, for example, super off-peak electricity rates (i.e., low nighttime power costs).

As noted, utility rates for any specific rate structure often vary by the season, with, for example, electricity more expensive in the summer than in the winter and with natural gas more expensive in the winter than in the summer.

Utility rates typically include both a generation rate and a transportation and distribution (T&D) rate, and some utility customers are allowed to obtain their utility (e.g., electricity, natural gas) from any available supplier. Therefore, for example, a facility can buy their natural gas from a wellhead supplier, or electricity from a remote power generator on the grid, and pay their local utility company for the T&D. In fact, some utility companies are trending toward being a T&D company and no longer generating or producing their own utilities. Regardless of from whom utilities are purchased, many utility bills show the division between generation charges and T&D charges.

Additionally, as noted, utility companies measure a facility's electric power use both by the amount of total electrical energy consumption (often measured in kilowatthours) over a particular time period (e.g., monthly) and by the peak power consumption rate, which is the maximum demand used (often measured in kilowatts) during the different utility rate time periods. Often, the peak demand is a separate charge on the utility bill. In some cases, the peak demand sets block of energy kWh costs, and sometimes a utility sets rates based on the annual peak demand, which may mean that the peak use in a single 15-minute period sets utility rates for an entire year. Often, a utility company measures, records, and charges for the maximum peak power demand used during each utility rate time period.

Utilities may have a number of energy charge and demand rate time periods such as, for example, an off-peak period (used for the nighttime period, weekends, holidays, or low-demand periods), a mid-peak period (typically for weekday mornings and early evenings), and an on-peak period (typically for weekday afternoons when peak utility demands occur, but can be for other periods of the day depending on when the time of peak demand is experienced by the utility company). Typically, the most expensive utility use and demand rates occur during the on-peak period, and the least expensive utility use and demand rates occur during the off-peak period.

Table 2-2 shows a typical time-of-use electric utility rate schedule. Note that there are both summer and winter rates and that during the winter period there is no on-peak period. There are different demand charges for the on-peak period and the mid-peak period as well as a non-time-related demand charge, which is applied to the facility's highest peak electric demand regardless of when that demand occurs, but it often occurs

	Non-time-related demand charge ($)	Summer		Winter	
		Demand charge, ($)	Energy charge ($)	Demand charge ($)	Energy charge ($)
On-peak		11.83	0.1223		
Mid-peak	2.85	4.51	0.0913	4.57	0.0861
Off-peak		—	0.0489	—	0.0423

TABLE 2-2   Typical Time-of-Use Electric Rate Schedule

during the on-peak period and therefore is added to the relatively high on-peak demand costs. Note that for the summer period, there is no on-peak period during the weekend or during holidays regardless of the time of day.

Effect of Rate Structures on the Most Cost-Effective Solution Chosen

The utility rate structure applied to a proposed thermal storage project can affect the results of which type of thermal storage system will be most economical. Key factors include the relative cost of utility demand charges and the ratio between on-peak and off-peak energy pricing. If there are no demand charges and no difference in energy costs between different periods of the day (i.e., no time-of-use rates), then partial thermal storage will likely be more economical than full storage systems because of smaller equipment requirements and smaller thermal storage tank requirements needed with a partial thermal storage system. However, if there are significant utility demand charges or if there are large relative differences between daytime utility costs and nighttime utility costs, then the full storage system may provide the greatest rate of return on investment.

New Investment Required to Meet New or Expanding Facility Loads

Sometimes, one alternative under consideration is to keep or maintain existing systems as is [the business-as-usual (BAU) case], and there is little or no capital cost associated with the BAU case. In this BAU case example, the capital costs for any other proposed alternative must be paid back from annual energy and maintenance cost savings at an acceptable rate of return on investment, which in some cases may be difficult depending upon relative capital cost requirements. On the other hand, when new facilities are planned or when facilities must be expanded to meet increased loads or even when old equipment must be replaced, then each alternative has associated capital cost. The various alternatives will (or should) be judged based on which alternative has the lowest lifecycle cost, that is which alternative has the lowest combined first cost and annual costs, taking into account the time value of money. This hurdle is true for any energy project, as it is typically always easier and more economical to incorporate green technology and energy conservation measures into new projects than it is to incorporate those technologies and measures into an existing functional system. Note, however, that greening existing facilities is an important challenge to be surmounted, since the vast majority of buildings and systems are existing.

Emergency Backup

Thermal storage can play an important role during an emergency. Some facilities have critical cooling (or even heating) loads, where even a small outage could be devastating. For example, a data center containing key financial or medical information requires continuous cooling and cannot tolerate a cooling outage even for 10 to 20 minutes while the cooling plant is brought back online. With some modest emergency power generation for chilled water pumping, thermal storage could be designed to provide 20 minutes of peak cooling to a data center, allowing the chiller plant to be brought back online during an outage. Likewise, a thermal storage system could provide emergency cooling to hospital surgery rooms and specialized equipment for a period of time during an outage while cooling equipment was made operational and operations or systems are safely shut down.

Fire Protection

Thermal storage systems, particularly aboveground stratified chilled water thermal storage tanks, have the potential to be incorporated into facility fire protection in a number of different manners; however, there are a number of important issues that must be addressed. As a minimum, it is fairly easy and inexpensive to incorporate fire department connections to a thermal storage tank for use as firewater in case of an emergency. Other thermal storage or fire protection options include the following:

- Use a code required fire water storage tank constructed to NFPA 22 standards as a thermal energy storage (TES) tank.
- Use a TES tank that is constructed to meet NFPA 22 as a main or backup firewater source.
- Connect TES tank to fire sprinkler system—in the United States, this must meet NFPA 13 Sec. 5-6.

In some cases, the building code may require (e.g., on high-rise structures), as part of the building construction, the addition of a firewater storage tank constructed to NFPA 22 standards. It may be possible to use that fire water storage tank as a TES tank, with additional costs potentially limited to just adding tank insulation, diffusers, temperature sensors, and some controls.

In the case where one is constructing a thermal storage tank to AWWA D100 standards, for example, constructing the TES tank to also meet NFPA 22 standards may be of minimal or marginal additional cost, involving, for example, using oiled sand for the tank base.

One of the key thermal storage fire protection incorporation or integration issues is the use of chemicals in a hydronic system for pH control, corrosion minimization, and biological control. The NFPA standard for the Installation of Sprinkler Systems, Sec. 5-6, Automatic Sprinkler Systems with Non-Fire Protection Connections, Part 5-6.1.8, Water Additives, states, "materials added to water shall not adversely affect the fire fighting properties of the water and shall be in conformity with any state or local health regulations." Common sense indicates avoiding any chemical additives that could be flammable or would harm the public or firefighters, considering also the effects to those chemicals by flame and heat exposure. One factor to consider is whether the TES system is always connected and part of the fire protection system or is the TES fire protection ability used only in a dire emergency.

Another TES or fire protection integration issue is the many differences in the requirements for a fire alarm system versus a plant energy management and control system (EMCS). Fire alarm requirements are typically governed by code and have specific requirements for providing backup power, loss of power notification, alarm notification, and for supervisory control (detection if the circuit has been broken, including a device not properly connected). EMCSs, typically, are not designed to a particular code (other than electric code and UL construction requirements, although building controls may, for example, control many code required functions such as providing minimum outside air quantities), and, as such, EMCS do not typically have more stringent and costly requirements. Therefore, integrating a TES system into a fire protection system may increase the TES system control costs.

Another issue is the allowed materials, connection types, and isolation requirements for a fire protection system versus a thermal storage system, and, when integrating the two, one must make sure that the fire protection code requirements and the thermal storage or hydronic system requirements are both met.

Additionally, there can be issues regarding testing of fire systems when they are coupled to thermal storage systems. For example, as hydronic system chemicals can be costly, flushing chilled or heated water down the drain for a sprinkler test would not be cost-effective. Periodic fire pump tests need to be able to be conducted without adversely affecting the thermal storage system capacity.

Any connection, incorporation, or integration of a thermal storage system with a fire protection system usually requires local building and safety fire department approval, and approval is, typically, handled on a case-by-case basis. Any proposed system, therefore, must meet both code and fire department requirements. The proposed thermal storage combination fire protection system may also need to meet facility insurance company requirements; however, meeting insurance company requirements may possibly lower insurance costs.

Combustion Turbine Inlet-Air Cooling

The net power output from a given CT (also known as a gas turbine) is a function of the inlet-air temperature to the CT. Nominal power output for a CT is typically rated at 59°F (15°C ISO conditions), and turbine power output is reduced anytime the outside air temperature is above 59°F, by a rule-of-thumb factor of approximately 1 percent for every 2°F increase in the outside air temperature. Therefore, for example, a CT operating at 100°F (41° above 59°F) would experience an approximate 20 percent power reduction from ISO conditions. The efficiency of a gas turbine is also reduced at higher air temperatures, and, therefore, the heat rate increases as more fuel is required to produce the same unit of power.

CT power plants are, therefore, known to exhibit power outputs and fuel efficiencies that are highly dependent on inlet-air temperature, and CT performance de-rating is known to be quite severe at elevated temperature conditions. In most locations and cases, it is precisely during those times of elevated ambient temperatures when power is most highly in demand and electricity has its greatest economic value.

The CT inlet air, however, can be cooled, for example, using chilled water in a coil, thereby increasing CT power output and increasing CT efficiency. The chilled water can be produced at night and stored in a thermal energy storage tank so that parasitic power for chilled water production will not substantially reduce daytime net peak electric power output. A particularly good application uses surplus steam in a steam turbine to produce additional electric power (combined cycle), and some of the power can be used during nights, when the electricity produced has less value, to operate electric-drive chillers with all or part of the cooling stored. The stored cooling can then be used during days when inlet-air temperatures are higher and the electric power generated has greater value.

Various means of CT power augmentation can be employed to increase power plant output. In particular, the family of CTIC technologies offers effective means of power augmentation during periods of hot weather and high-value power. More specifically, in those hot weather conditions, CTIC accomplished using chiller technology is well developed and is known to provide (1) dramatic increases in hot weather CT power output; and (2) significant reductions in installed capital cost per unit of power output.

Through the integration of thermal storage technology with CTIC, further advantages are often gained in both plant performance and project economics. As thermal storage decouples the generation of cooling from the use of cooling, thermal storage can be configured to significantly reduce, or even nearly eliminate, the parasitic power consumption of the CTIC system during the peak power demand periods of the day. Thus, net power output is maximized during those critical times. Furthermore, the addition of thermal storage can effect a large reduction in the capacity and cost of the installed chilling technology, and, therefore, net capital cost for the CTIC system can be reduced. In this way, the installed unit capital cost per MW of power enhancement, already attractively low for conventional (nonthermal storage) CTIC systems, can be reduced even further with thermal storage.

There are many examples of actual CTIC-thermal storage installations, and they cover a broad range of applications such as the following:

- Size—from 1 to 1000 MW
- Plant type—including simple cycle, combined cycle, and cogeneration
- Project timing—including new construction and retrofits
- Locale—on at least four continents
- Weather conditions—including hot-humid and hot-arid climates as well as in year-round hot climates and in highly seasonal climates
- Owner or operator type—including power utilities, IPPs, District Energy, and even LNG import terminals (using recovered cold from NG revaporization)

As noted, CT power plants are well known to suffer from the adverse impact of high inlet-air temperature, in reduced power output (as well as reduced efficiency). Moreover, unfortunately, the periods of highest ambient air temperatures (and greatest de-rating of performance) are generally precisely those times when power demand is highest and power output is most valuable. Accordingly, CTIC technologies of various types are increasingly being employed to cool the inlet air, and thus enhance the power output (and improve the heat rate), of CT power plants. This is being accomplished in a wide range of applications, including the following:

- New CT plants and retrofits to existing CT plants; for industrial and aero-derivative CTs
- Simple-cycle and combined-cycle CT plant configurations
- Utility, independent power, and distributed generation (DG) plants
- A range of plant capacities (from 1 to 1000 MW); in locations around the world
- A range of climates (from hot-arid to hot-humid and from hot year-round to seasonal)

The leading approaches for CTIC include two families of technologies:

1. Evaporative cooling methods (which have low initial cost but can cool only the inlet air to near the ambient wet-bulb temperature)

2. Chiller-based systems (which have a higher initial cost but typically cool the inlet air to well below the ISO conditions, for example, to 40 to 50°F (4 to 10°C), and achieve a net power enhancement in hot weather of 20 to 30 percent or more

Table 2-3 presents a summary of data from a survey of CT plants employing CTIC with thermal storage. Data are presented for several dozen CTs using thermal storage-CTIC at more than one dozen power plants in North America, Europe, the Middle East, and Southeast Asia. Statistics are presented to illustrate the locations, ages, types, and sizes of plants; the thermal storage technology types; the level of power enhancement; and trends over time.

Numerous examples of thermal storage applications exist, documenting the actual benefits achieved by facility users, both in CTIC power augmentation and in the much more numerous traditional demand-side management (DSM) applications. The variable time-of-day nature, of the demand for and the value of power, is driving increased deployment of thermal storage in these applications. In addition, in large applications, thermal storage is being increasingly used for its ability to reduce capital investment relative to that for equivalent conventional (nonthermal storage) chiller plant capacity.

Similarly, thermal storage is also realizing increasing application in the power generation market, where it is used by utilities and by on-site power generators for CTIC. Some specific examples of thermal storage-CTIC systems are listed in Table 2-3, along with key application data. Hot weather power augmentation ranges from 16 to 42 percent, with most applications being 21 to 35 percent.

In comparing documented cases in earlier installations with more recent applications, some strong evolutionary trends are apparent including trends from the following:

- Strictly utility or cogeneration plants to utility, cogeneration, district cooling, hospital, and university plants

- Primarily U.S. locales to U.S. plus worldwide locales (on four continents)

- Medium-scale CTs to a broad range of scales (1 to 75 MW CTs; 3 to 750 MW plants)

- Latent-heat thermal storage to sensible-heat thermal storage

- Primarily ice harvester thermal storage to stratified chilled water or low-temperature fluid thermal storage

- Short on-peak design periods (4 to 6 h/day) to longer periods (6 to 10 h/day)

- Weekly cycle thermal storage (charge nights plus weekends) to daily cycle configurations

- Single-use cooling (dedicated to CTIC) to also include dual-use (CTIC + other cooling)

Many issues in the current electric power marketplace mandate the need to optimize asset value, specifically through the consideration and implementation of power

Year	Turbine power plant Application–location	CT quantity Make & model	Thermal storage Type	Thermal storage (Ton-hours)	Capacity (h/d)	Power Boost (%)
1988–1997						
1988	Cogen–California	GE LM 2500	CHW	3,500	6	16
1991	Utility–Nebraska	GE 7B	Ice harv'r	45,000	4	27
1993	Utility–North Carolina	8 × GE Frame 5	Ice harv'r	not avail.	4	26
1994	Cogen–California	GE 7EA	Ice harv'r	40,000	6	24
199x	Cogen–California	CT	Ice harv'r	14,800		23
1995	Utility–Malaysia	6 × GE 7EA	Ice harv'r	not avail.	4	28
1995	Utility–North Carolina	W 251	Ice harv'r	not avail.	4	42
1997	Utility–Nebraska	2 × CT	Ice harv'r	165,000		21
Subtotals for 10 years (1988–1997)				429,280 (total)		
8 examples of thermal storage-CTIC	21 CTs in 7 of 8		Ice harv'r	53,660 (avg.)	4.7	26
1997–2005						
1997	District Cooling–Illinois	3 × Turbomeca	LTF	123,000*	13	35
1998	District Cooling–Florida	GE LM 5000	CHW	57,000*	10	31
1998	District Cooling–Portugal	CT	CHW	39,800*	10	17
1998	Utility–Australia	3 × GE Frame 6	Ice-on-coil	36,932	4	21

TABLE 2-3 Examples of Thermal Storage Used for Turbine Inlet Cooling

| Year | Turbine power plant | | CT quantity | Thermal storage | | Capacity | Power |
	Application—location		Make & model	Type	(Ton-hours)	(h/d)	Boost (%)
1998	Utility—Saudi Arabia		6 × GE 7EA	Ice harv'r	120,000	5	35
1999	Cogen—Texas		3 × W 501 D5	CHW	107,000	10	21
2005	Hospital/DC—Texas		Solar Mercury 50	CHW(LTF)	8,000*	6	
2005	University—New Jersey		GE LM 1600	LTF	40,000*	4–10	
2005	Utility—Saudi Arabia		10 × GE 7EA	CHW	193,000	6	30
Subtotals for 9 years (1997–2005)				CHW/LTF	724,732 (total)		
9 examples of thermal storage-CTIC			29 CTs in 7 of 9		80,526 (avg.)	7.9	27
Totals for 18 years (1988–2005)					1,154,012 (total)		
17 examples of thermal storage-CTIC			50 CTs various		67,883 (avg.)	6.4	27

* indicates thermal storage capacity is used for CTIC but also for other uses, namely district cooling.

TABLE 2-3 Examples of Thermal Storage Used for Turbine Inlet Cooling (*continued*)

augmentation. For CT-based power plants, such augmentation can be accomplished through any of various types of CTIC. An often particularly advantageous approach to CTIC involves the combination of chiller-based cooling with thermal storage.

The use of thermal storage-CTIC is applicable in new and existing CT-based power plants, in simple- and combined-cycle installations, in a wide range of sizes, in any hot weather locale, and in applications including utility power generation, merchant plants, cogeneration or CHP-combined heat and power, and on-site or DG applications (such as hospital, university, and district cooling applications).

Thermal storage, properly applied, can both minimize the capital cost of chiller-based CTIC and maximize the net power output enhancement of CTIC during high-demand, high-value, on-peak periods. Thus, thermal storage-CTIC can often deliver not only rapid payback and high returns on investment but also the maximum net present value of various power augmentation options. Thermal storage-CTIC can be expected to have continuing increases in its application.

CHAPTER 3
Types of Thermal Storage Systems

As discussed in this book, thermal storage systems include any process that can produce heating or cooling at one point in time, store the product, and then use the product at another later point in time. The thermal production can be a natural process, a mechanical process, or a chemical process. Thermal storage provides either heating or cooling for a building, a group of buildings, manufacturing, or other processes. Note that it is somewhat an anachronism to call any process used for cooling, "thermal storage." Chilled water (CHW) can indeed be stored, but storing CHW is not storing energy, it is storing the potential to accept heat from something warmer. In the case of heating thermal storage, a process is used to heat a material to achieve thermal storage and the product (e.g., hot water supply) becomes cooler with use. A cooling thermal storage process is cooled to achieve storage and becomes warmer with use.

There are many different types of thermal storage systems, which are divided into two broad groups: sensible thermal storage systems and latent thermal storage systems. Sensible thermal storage systems achieve thermal energy storage by a difference in a material or substance's temperature, whereas latent thermal storage systems achieve thermal energy storage by a change in a material or substance's phase (e.g., water to ice or ice to water). Thermal storage systems are further characterized by the following:

- Whether the system is active or passive
- Whether the purpose is for heating or cooling
- The medium used for thermal storage
- The thermal storage cycle period (e.g., daily, weekly, seasonal, or emergency)
- The thermal storage operating strategy (e.g., partial storage versus full storage)
- Whether the thermal storage system is above or below ground

Active thermal storage systems use mechanical energy and transport to achieve thermal storage, whereas passive thermal storage systems do not. Examples of passive thermal storage systems include high mass walls exposed to sunlight designed to allow natural convection heat transfer to the building. Some thermal storage applications are produced by nature without machinery, such as naturally produced cold water in a lake in winter or the high mass wall heated by the sun. Naturally produced cooling or heating storage may require mechanical systems such as pumps or fans to distribute thermal energy, or the thermal storage systems may naturally heat

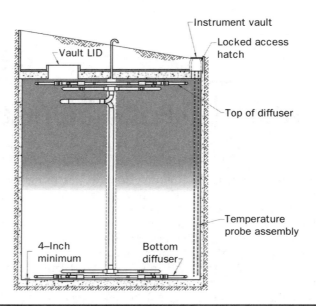

Figure 3-1 Belowground stratified chilled water tank section.

or cool by convection or radiation. Examples include a northern U.S. university that draws on cold lake water produced naturally during the winter to cool its campus in the summer. In that case, nature provides both the cooling water and the thermal storage vessel, but mechanical systems (piping, pumps, heat exchangers, etc.) are needed to use the natural storage. Similarly, in Hawaii, a district cooling system (under development in 2011) is able to draw cold deep water that is cooled by nature; however, mechanical systems and energy are required to collect and transport the cold ocean water. Another example might be a gravel bed, using naturally produced cold outside air forced through the gravel bed by fans to achieve thermal storage. Most thermal storage systems in use today require machinery and electrical or chemical energy to produce the stored heating and cooling as well as mechanical systems to deliver and use the stored thermal product.

Although there are some impressive large-scale seasonal thermal storage systems, the majority of facilities and buildings use a daily, 24-hour, thermal storage cycle. In addition, while there have been various thermal storage materials used as the thermal storage medium, the most common thermal storage material is water for the reasons outlined in Chap. 1.

Therefore, for example, one type of thermal storage system might be characterized as an aboveground, stratified CHW thermal storage system designed as a daily cycle full storage system to reduce peak facility demand associated with cooling loads to the maximum extent possible. And another type of thermal storage system might be characterized as a belowground, ice-harvester thermal storage system designed as a weekly cycle partial storage system to meet Sunday cooling loads at a church.

Each type of thermal storage system has its own general advantages and disadvantages with regard to the following:

- Cost to construct (capital cost)
- Overall efficiencies
- Annual energy costs
- Operation requirements
- Maintenance costs
- Required equipment types
- Space required
- System longevity
- Toxicity and other environmental issues
- Visual issues
- Engineering needed

However, specific advantages and disadvantages can and do vary from project to project and case by case depending upon a number of factors, for example, the size of the load served, the space available for thermal storage, the availability of electric power, and whether the project is new or an existing construction.

Sensible Thermal Storage Systems

As discussed, sensible thermal storage systems achieve thermal storage by a change in the temperature of a substance or material. When energy is added to a material, the temperature of the material rises. Likewise, when energy is taken from a material, the temperature of the material drops. The amount of temperature change depends on the mass of the material, the amount of energy added or taken away, and the specific heat of the material. Every material has its own specific heat constant. The larger the specific heat constant, the lesser the mass that is needed for a given amount of thermal storage. Additionally, a heavy (denser) material has more mass per unit volume, and so a heavy material can provide more thermal storage than the same volume of a lighter, less dense, material with the same specific heat constant. The amount of sensible thermal storage depends on both the mass and the specific heat constant of the material.

Another factor affecting sensible thermal storage systems that use solid materials as the thermal storage medium is the storage material's thermal conductivity (ability to transfer heat) and the thickness of the material to its center. Thermal conductivity is usually expressed as a steady state conductance. In the English system of units, the unit for thermal conductivity is British thermal units-per hour-per-foot-per-degree Fahrenheit of temperature difference (Btu/h/ft-°F). Steady state is only achieved with a solid material after a period of time. When the steady state heat transfer is achieved, there is a temperature difference from the solid's exterior surface to the center of the material. When heat is applied from the exterior, over time, the center of the solid rises in temperature until the entire solid is one temperature. The result of this phenomenon is that a solid material degrades in the rate of energy transferred into or out of the storage system over time. The rate of energy transferred is greatest at the beginning of a thermal storage cycle, and the rate of energy transferred becomes less until the entire storage material is one temperature, at which point, all heat transfer stops.

An additional heat transfer factor is the surface film between the solid storage material and the transfer medium used to heat or to cool the storage material. Most often this media is either water or air. A still (i.e., nonflowing) water or air film reduces the heat exchange rate between the transfer medium and the solid storage material. Liquid storage materials do not have this characteristic and are usually heated or cooled directly and then transferred to a storage vessel(s) of some type.

Sensible thermal storage systems include hot water (or hot oil, or molten salt) thermal storage, CHW (or other chilled fluid) thermal storage, some types of seasonal thermal storage, and mass thermal storage, which includes the use of gravel, earth, building mass (concrete and steel), and ceramic bricks.

From thermodynamics, the amount of energy required to raise a material's temperature (i.e., sensible heat) is given by the formula:

$$Q = m^*c^*(T_h - T_c)$$

where:

Q = the quantity of heat
m = the mass of the material
c = the specific heat constant for the particular material
T_h = the hot temperature, or the final temperature
T_c = the cold temperature, or the initial temperature

It should be noted that $T_h - T_c$ is usually called delta-T for change or difference in temperature.

The specific heat constant, c, as noted, varies from material to material and is equal to the amount of heat required to raise one unit of mass 1° in temperature. For common thermal storage materials, specific heats are typically constant over the temperature ranges seen in most thermal storage systems. In English units, by definition, water has a specific heat value of 1 Btu (British thermal unit) per pound-mass (lbm) per degree Fahrenheit (F). So for example, 50 Btu are needed to raise 5 pounds (lb) of water at 10°F. In metric units, the specific heat constant for water, by definition, is 1 calorie per gram (cal/g) per °C . A list of specific heat constants for various materials can be found on the Internet or in any heat transfer, thermodynamic, or physics book; Table 3-1 provides the thermal properties of some common thermal storage materials. Also, note that, using calculus, the previous formula can be differentiated with respect to time so that the heat transfer rate (e.g., Btu-per-hour) is equal to the mass flow rate times the specific heat constant times the delta-T.

Hot Water Thermal Storage Systems

Hot water storage is one of the oldest types of thermal storage systems, and is also one of the most common types of thermal storage systems in existence today, as most modern houses have a domestic hot water storage tank. In addition, many parts of the world employ rooftop solar hot water heating, which use thermal storage. In addition to domestic hot water storage tanks and solar hot water storage tanks (which can be a

Material	Specific heat	Density
Water	0.49 Btu/lbm-°F @ 32°F (ice) 1.0 Btu/lbm-°F @ 32.01°F 0.999 Btu/lbm-°F @ 70°F 0.998 Btu/lbm-°F @ 95°F 1.0 Btu/lbm-°F @ 212°F	57.2 lbm/ft^3 @ 32°F (ice) 62.4 lbm/ft^3 @ 32.01°F 62.3 lbm/ft^3 @ 70°F 62.1 lbm/ft^3 @ 95°F 59.8 lbm/ft^3 @ 212°F
Gravel	0.18 Btu/lbm- to 0.24 Btu/lbm-°F	105 lbm/ft^3
Concrete	0.2 Btu/lbm-°F	133 lbm/ft^3
Earth	0.19 Btu/lbm- to 0.21 Btu/lbm-°F	78 lbm/ft^3 to 125 lbm/ft^3
Eutectic salt	50 Btu/lbm-°F	91 lbm/ft^3

TABLE 3-1   Thermal Properties of Common Thermal Storage Materials

variation of a DHW storage tank), hot water for space heating can also be stored in thermal storage tanks. Hot water thermal storage can be a good fit with heat pump systems, where, instead of rejecting heat to the environment during cooling production, that heat can be captured and stored as hot water in a thermal storage tank for later use during the evening and the following morning, for example.

Hot water has always been a simpler thermal storage process to achieve than CHW storage because of several factors. First, there is usually a much greater temperature difference in hot water storage system between the cold and hot waters in the storage tank. Second, the density change per degree of temperature difference for water graph has the shape of a hyperbolic function, which approaches zero density change as it nears the maximum density point for water near 39°F. In hot water systems, the large density difference, due to the temperature difference between the hot water supply and the cold water properly introduced into the thermal storage tank, provides significant natural thermal stratification. With CHW at or near the maximum density point, the density difference effect due to fluid temperature difference is much less powerful.

Hot water used with thermal storage can be generated from a number of sources, including solar energy, various fuel-fired boilers (including biomass fired boilers), CHP plant heat recovery systems (waste heat boilers), heat pumps, air compressors, and refrigeration or chiller equipment heat rejection. For example, a southern California university recovers heat at 149°F from a 1250-ton central plant chiller during the peak of the day and stores that recovered heat in a hot water thermal storage tank for use later at night and the following early morning.

Hot water thermal storage, as any thermal storage system, can allow for smaller hot water generation equipment than with a conventional plant heating system and is likewise sized based on the load profile and hours of heat generation. Similarly, hot water thermal storage also provides a method to decouple when heat is generated and when that heat is used, which can be important when employed with CHP (cogeneration) systems or solar water heating.

Hot water thermal storage tanks may be roof-mounted or ground mounted, used for a variety of applications from residential to industrial, constructed from steel, plastic, or concrete, and are typically installed in a vertical orientation. During discharge or use, hot water is drawn from the top of the hot water thermal storage tank, and cold water enters the bottom of the tank. During the charging mode, cold water is removed or pumped from the bottom of the tank and is heated by the heating source and returned to the top of the hot water thermal storage tank.

Solar Water Heating

Solar water heating is a mature renewable energy technology and is widely used in China, Australia, Japan, the European Union, including Austria, and around the Mediterranean, India, and Israel. Additionally, solar water heating has been used for more than 100 years in the United States. Virtually all solar water heating systems use some type of thermal storage tank to store hot water produced in the peak of the day for later use at night and the following morning.

Solar water heating systems typically fall into two broad categories: active and passive systems. Passive solar heating systems, sometimes called compact or close-coupled systems, work by natural convection, and no pump is needed to circulate fluids. With passive solar systems, the thermal storage tank is located above the solar collector since, as noted, hot water has a tendency to rise and cold water has a tendency to fall due to the density difference. Passive solar water heating systems tend to be limited to smaller systems and are presently most often used for residential systems.

Active solar heating systems have a pump to circulate the fluid; and the thermal storage tank can be mounted on the ground below the solar collectors, for example. Active solar heating systems are capable of achieving higher delta-Ts than passive solar heating systems; but active systems are more complicated, requiring pumps and controls, and are typically more costly than passive systems. The ability to locate a thermal storage tank on grade may have advantages versus a roof-mounted tank (e.g., potentially less structural requirements and subsequent costs).

Solar water heating often works best in hot climates, where solar radiation is stronger than in other latitudes and where there is less heat loss. There are a number of different types of solar collectors including the following: flat plate, integral collector storage system (ICS), evacuated tube, and parabolic mirror troughs. Most solar water heating systems require backup hot water generators or boilers to provide heat during cloudy periods or inclement weather, during the winter when there is less solar radiation, and to boost the hot water supply temperature to the required temperature at the use point (when the solar water heating system may not be able to reach the required final hot water temperature). Conditions that reduce or eliminate solar water heating production require a larger thermal storage tank to help account for some of these conditions.

Solar hot water systems may be direct systems, where the hot water is used directly by occupants, or indirect systems that use a separate solar heating loop with a heat exchanger(s) to heat domestic hot water, for example. With an active, indirect solar heating system, domestic cold water (DCW) is pumped from the bottom of the thermal storage tank through a heat exchanger, where the DCW is heated by the solar heated water loop, and back into the top of the storage tank. In all cases, the thermal storage tank needs to be well insulated to minimize the heat loss rate, and many solar water heating systems have a drain-back tank to empty the solar collectors at night or during cold weather conditions.

Solar water heating systems are a particularly good fit for swimming pool heat and for domestic hot water production, where a well-designed system may provide for more than 80 percent of a facility's annual domestic hot water loads.

Chilled Water Systems

CHW thermal storage systems are also a common type of thermal storage system, and have been in use for almost four decades. CHW thermal storage systems have proven to be highly successful and offer some advantages over competing thermal storage strategies, but they also have some shortcomings. CHW thermal energy storage systems are custom built, and, thus, require individual project engineering design. In particular, successful CHW storage systems exhibit some necessary or highly desirable design considerations such as the need for a constant, large CHW delta-T, variable flow, and the need for pressure control that are not commonly applied on nonstorage HVAC systems and that may not be necessary on other types of thermal storage systems. There are also a few different types of chilled thermal storage tank configurations, though today the most common type of CHW thermal storage tank is the thermally stratified CHW tank.

Like hot water storage systems, CHW storage systems are sensible heat systems in which the thermal storage is achieved by a change of water temperature. Typical CHW storage systems are usually designed using a 15 to 30°F temperature range between CHW supply and CHW return temperatures. Using a 20°F temperature range, CHW can provide about 1200 Btu/ft³ of thermal storage. This relatively lower value means that about three to four times the storage volume is required with CHW thermal storage compared with ice or phase change materials (note, however, not all of the water is typically turned to ice).

Chilled thermal storage systems typically use chillers to generate CHW, pumps for distribution, and either an aboveground insulated thermal storage water tank(s) or belowground thermal water storage tank(s) that may or may not need insulation depending upon tank construction and soil temperatures. Cold CHW supply is supplied to or withdrawn from the bottom of the thermal storage tank via specially designed flow diffusers, and "warm" CHW return is returned to or withdrawn from the top of the thermal storage tank via flow diffusers. The tank diffusers are designed so that all water introduced into or withdrawn from the thermal storage tank is done so in a laminar fashion (even at the peak design flow rate) to prevent any unwanted mixing due to turbulence or eddies of water through the thermal storage tank.

As the thermal storage capacity of a CHW system is directly proportional to the CHW delta-T (i.e., the difference between the CHW return temperature and the CHW supply temperature), the magnitude of the CHW delta-T is critical to the sizing and economics of a CHW thermal storage system. The importance of delta-T and its effects on sensible thermal storage systems is discussed further in Chap. 4.

All CHW thermal storage systems must carefully separate the warm CHW return from the cold CHW supply to prevent mixing, which would degrade the thermal storage capacity. There are a number of concepts to separate the CHW return from the CHW supply including the following:

- Vertical thermal stratification
- Membrane separation
- Labyrinth design
- Empty tank (multiple tank)
- Series tank

Each separation concept leads to a somewhat different type of CHW thermal storage system. However, as noted, vertical thermal stratification is the most common type of sensible thermal storage system in use today.

Three of the above separation concepts operate on the same principles but obtain separation in a different way. These are vertical stratification, membrane separation, and labyrinth design. A fourth system, empty tank, operates on an entirely different principle.

One type of empty tank thermal storage concept uses a number of individual thermal storage tanks piped in series so that the overflow from one tank can travel to the next tank with an empty tank acting as an interface between the CHW supply and CHW return. The empty tank "moves" up and down the line of thermal storage tanks as the system is charged and discharged.

As previously discussed, the vertical thermal stratification system depends on the difference in density between warmer CHW return and colder CHW supply. This density difference, although small, is enough to ensure that if water is carefully introduced via a diffuser system into a tank with warmer, more buoyant water on top, it will remain in the same relationship to the colder more dense water below it. CHW must be moved through a tank at low velocities in order not to disturb the relative position of warmer water on top and colder water below. However, even with low water velocities, some mixing (due to mass transfer and heat transfer) occurs at the interface between the cold CHW supply and the warm CHW return and creates a zone in the thermal storage tank where the temperature changes from cold supply water temperature to warm return water temperature. In the case of a stratified storage system, this temperature transition zone is called the thermocline and can be less than 3 feet thick with good diffuser designs. The thermocline thickness represents unusable thermal storage water volume, and in taller thermal storage tanks represents a smaller percentage of unusable volume than in shorter thermal storage tanks.

Membrane separation systems use a single tank with a diaphragm to separate the CHW return and CHW supply. The diaphragm moves up and down using the tank in a way that may be compared to the moving empty tank. The membrane separation design concept was offered as a solution to the interface problem, it does not, however, eliminate the entire problem, since heat transfer still occurs across the noninsulated membrane producing a thermocline. Additionally, the diaphragm membrane has been prone to mechanical failure. Membrane systems are not typically used in today's market.

Series (or "baffle and weir") tank designs are multiple tanks (or tank sections) arranged with the bottom of one tank connected to the top of another. This arrangement creates all of the thermal stratification advantages as a single tank as tall as the sum of all the tanks. A labyrinth solution is a design of many tanks arranged in a way that water from one tank flows to the next and is similar to the series tanks described above. The two primary differences between series and labyrinth tank arrangements are the number of tanks and the locations of connections.

Advantages of Chilled Water Storage Systems

The following are several general observations on the advantages of a CHW thermal storage system contrasted with other potential thermal storage systems. In certain cases, the advantages may not apply compared to all other types of thermal storage systems.

Equipment used   A typical CHW thermal storage system uses the same pumps, chillers, and other components of the cooling system that a conventional CHW cooling system uses. Often a CHW thermal storage system benefits from producing colder CHW than conventional chilled systems. Controls are often installed as part of the thermal storage system to assure a constant high CHW return temperature.

In the case of retrofit designs or expansion of existing systems, the ability to use the existing chiller and other related systems is an important benefit of CHW systems and can dramatically reduce the storage system costs.

System life A CHW thermal storage system typically provides a long machinery life as the system ordinarily uses long life, centrifugal water chillers and pumps, which normally last 20 to 30 years (note that some ice systems also use centrifugal chillers and have long lasting storage tanks). Even though chillers may operate more hours a day, that does not ordinarily reduce the equipment life since the chillers experience fewer overall stops and starts. The use of reciprocating equipment with its 12- to 15-year life is normally avoided with CHW thermal storage systems. The water-filled, no moving parts, thermal storage tanks normally require little or no routine maintenance and can last for many decades.

Economics Every potential thermal storage situation is different, and these differences may affect thermal storage feasibility of any specific case; the only reasonable way to determine the best answer is to study the alternatives. Because of the differences, it is dangerous to draw too heavily on generalities. Nevertheless, the conclusions presented herein are based on the study of over 70 potential thermal storage systems in all regions of North America. From those studies some general information can be derived about the patterns that have developed.

Size is one of the major factors affecting the cost of CHW storage systems in comparison with other solutions. Small systems, less than a few thousand ton-hours, are typically better economically suited to packaged ice (or other package) solutions than to larger CHW systems, which require individual engineering design. Additionally, there is a major cost reduction in thermal storage costs per unit volume with larger CHW storage systems.

For systems larger than 5000 ton-hours of thermal storage, CHW storage systems may show the lowest overall ownership costs when system life, energy costs, first costs, maintenance, and the time value of money are taken into account (i.e., life-cycle costs). In some cases, a competing solution may show lower first cost, but ordinarily that lower first cost is achieved by using lower cost components that may not have a long life and are not able to provide the operational energy efficiency and economics of a CHW storage solution.

Material cost Water is one of the more common materials available, provides good sensible heat and latent heat characteristics, does not wear out, and is simple to handle, and nonhazardous. In evaluating the storage systems, the cost of the tank or container, the pumps, the chillers, and other materials represent nearly the entire thermal storage system cost. With environmental concerns rising, the use of glycol or salts may become subject to stringent rules.

Disadvantages of Chilled Water Thermal Storage
CHW thermal storage also has disadvantages in comparison with other thermal storage systems. The following are general observations of the disadvantages of a CHW storage system in comparison with other potential thermal storage systems. In certain

cases, the disadvantages may not apply when compared to other types of thermal storage systems.

Larger storage volume   When a knowledgeable designer compares storage systems, the greatest disadvantage usually noted is the greater storage volume required of CHW thermal storage systems versus other storage systems. In general, a water storage system may require up to three to five times the volume of an ice or eutectic solution. The required storage volume can be reduced by increasing the delta-T.

Pressure control   Because of the high cost of constructing pressure-rated tanks in the size needed, most thermal storage systems ordinarily use nonpressurized, atmospheric tanks. This means that for tall buildings, some form of pressure control is necessary to interface with the pressure in the building CHW piping system when using the nonpressurized storage tanks. Most CHW thermal storage systems share this requirement; however, most ice systems, for example, use heat exchangers, which eliminate the problem. Typically, backpressure control valves (also called pressure sustaining valves), backpressure regenerators (recovery turbines), heat exchangers, and gravity return systems are other solutions used to solve the pressurization problem. When water storage systems incorporate tanks with a water level above the height of the buildings, this eliminates the problem.

Special design requirements   As noted, CHW storage systems have special requirements and design considerations that are not normally considered in traditional ice or eutectic system design. These items include the following:

- Need for high temperature rise in cooling coils; and variable CHW flow in cooling coils or some other method of obtaining full design temperature rise to the CHW under various load conditions
- A separation method between CHW supply and CHW return and
- Coil load two-way control valves that close when there is no air flow across the cooling coils.
- These and other design requirements are discussed further in Chap. 4.

Mass Storage Systems

Sensible thermal storage can be obtained in the temperature change of almost any material. Many thermal storage systems have been built using gravel. Gravel has the advantage of being able to produce air temperatures very close to the storage temperature because of the large area of thermal contact. The systems in some cases can be relatively inexpensive. Systems using gravel typically use cooler night air and evaporative cooling for production of cooling. In some cases, air quality problems may result, including damp and musty odors. In certain cases, tests have shown that gravel systems require more energy than conventional cooling solutions.

Concrete or other building materials are sometimes used to provide at least part of the building cooling during on-peak energy periods by subcooling a building during the night or outside the peak electric or peak cooling demand period, with large theaters, auditoriums, or sports arenas being common examples. The idea is as old as buildings themselves. In areas with substantial daily temperature swings, the windows are often opened. In other cases, buildings are sub-cooled to reduce the load at a peak period. These concepts affect the peak loads but should not be considered to be true thermal storage systems. Because a true thermal system is reversible and controlled, it does not overcool the building when cooling is not needed. When buildings are air conditioned and fully occupied around the clock, the only variable is the outdoor gain factors. The mass of the buildings averages the effect of changes in outside air temperature and solar gains.

Electric thermal storage (ETS) heating systems (or ceramic brick thermal storage) are another form of a sensible thermal mass storage device, which are used in cold weather areas. ETS systems use electric resistance heat at night during the low-cost off-peak period to heat up high-density ceramic bricks located in a well-insulated box. When heat is needed the following morning or during the day, one method uses a fan to blow air over the ceramic bricks to transfer heat from the bricks to the air for distribution to the facility or building. ETS systems can also be used with hydronic systems where fluid heated from the hot bricks is pumped to heat exchanger coils (air-handling units, fan coil units, reheat coils, and process heating coils). ETS systems can cost only half as much to operate as a real-time electric heat system without thermal storage. ETS systems have some challenges including energy loss (though the ETS system can be kept inside an occupied space so that any heat loss goes to beneficial use); lack of flexibility as the heat will be discharged the following day either directly to space uses or indirectly as heat loss, whether the heat is needed or not (e.g., residential occupants may unexpectedly be away for the day); and the fact that ETS systems run out of heat storage in the evening when people most often want heat. Fuel costs, time-of-use electricity rates, and utility rebates are key factors in the economic applicability of the ETS systems.

Latent Thermal Storage

Unlike sensible thermal storage systems, latent thermal storage systems achieve their thermal storage by a change in phase, typically either from solid to liquid or liquid to solid. Heat is absorbed when a solid melts and, conversely, heat is released or removed when a liquid solidifies. Each material has its own heat of fusion (the amount of energy required per unit of mass to cause a phase change, when at the phase change temperature). Table 3-2 shows some heat of fusion values for various materials. Latent thermal storage systems include ice systems, eutectic salts, and other phase change materials; however, the majority of latent thermal storage systems today are ice storage systems, as water is an excellent thermal storage medium, as previously discussed, and has a high latent heat of fusion (144 Btu/lbm) compared with other materials and changes phase at 32°F. In practice, ice thermal storage systems can store about 5000 Btu/ft^3 depending upon the type of ice storage system (as noted, not all of the water typically turns to ice).

Ice Systems

There are a number of advantages associated with the use of ice storage systems including possible

- Water distribution savings due to a higher system delta-T
- Air distribution savings due to the use of cold air distribution
- Minimal water treatment required with a glycol-based systems

Material	Melting temperature	Heat of fusion
Water	32°F	144 Btu/lbm
Eutectic salt*	47°F (typical)	41 Btu/lbm
Paraffin waxes	Melting points for paraffin waxes have a range of 125–165°F and may be classified into three general grades: Low Melting: 125–135°F Mid Melting: 135–145°F High Melting: 150–165°F	86–95 Btu/lbm

* There are a variety of eutectic salts with different melting points and heats of fusion.

TABLE 3-2   Heat of Fusion for Common Thermal Storage Materials

with a major advantage in the fact that ice systems are not affected by low delta-T syndrome the way sensible thermal storage systems are affected. Some disadvantages of ice storage can include the need for ice-making equipment and the higher energy required to make ice versus CHW (although this energy penalty may be offset as discussed in Chap. 5).

There are a number of types of ice thermal storage systems including the following:

- External melt ice-on-coil systems
- Internal melt ice-on-coil systems
- Encapsulated ice
- Ice harvesters
- Ice-slurry systems

Each type of ice thermal storage system has its own advantages and disadvantages.

Some ice thermal storage systems, such as internal melt (or external melt) ice-on-coil systems, use modular, pre-engineered, packaged thermal storage tanks, which can be grouped together in parallel to achieve virtually any thermal storage capacity (within reason). Other ice thermal storage systems, such as encapsulated ice (or external melt ice-on-coil), often use a single engineered thermal storage tank.

The term "internal melt" versus "external melt" refers to whether the ice (which has been previously formed on the outside of an internally cooled coil of piping or tubing) is melted from the inside (internal melt) or whether the ice is melted from the outside (external melt). With an external melt ice-on-coil thermal storage system, heat exchanger coils flowing cold refrigerant or chilled water-glycol through the inside of coils forms ice on the outside of the coils during the charging period. During the discharge period, warm system CHW return flows in direct contact over the outside of the ice, in the process melting the ice from the outside (external) and thereby reducing the CHW temperature creating CHW supply. With an internal melt system, the charging process is the same as with an external melt system; however, the discharge process is different.

During the discharge period, the warm CHW return (typically a water-glycol-brine mixture) flows inside the heat exchanger tubes where it is cooled, and in the process the ice is melted from the inside (which is the outside of the heat exchanger tube). Ice-on-coil systems have a storage capacity of approximately 5000 Btu/ft^3.

External Melt Ice-on-Coil Thermal Storage Systems

External melt ice-on-coil storage systems are one of the oldest types of ice storage systems, which are used as a refrigerated coil inside a tank of water to produce ice. Instead of a refrigerant circuit within the coils, a water-glycol mixture can be used as the coolant, which can be produced directly in a low-temperature chiller or via a refrigerant-to-water heat exchanger. With an external melt system, the water in the tank is never completely turned to ice, since the water is the system coolant for the building or facility, which is pumped from the thermal storage tank to facility air-conditioning and process loads. One challenge with external (or internal) melt ice-on-coil systems is that as the ice grows thicker around the coils during the charging process, heat transfer is reduced (ice is an excellent thermal insulator), and hence ice production becomes more difficult and ice-making equipment must work harder (which takes more energy) per unit of cooling produced. With external melt systems, controls are provided to measure the buildup of ice to ensure that there is no ice "bridging" between adjacent coils (which would block the necessary external water flow during subsequent melt cycles). Because of the direct contact of the CHW and the ice, advantage of external melt ice thermal storage systems include the ability to achieve low temperatures and high discharge rates.

Internal Melt Ice-on-Coil Thermal Storage Systems

Internal melt ice storage systems are one of the most common commercially available ice thermal storage systems today and can be obtained from a number of manufacturers (see Fig. 3-2). With an internal melt ice storage system, standard chillers capable of producing low-temperature chilled water-glycol mixture at about 25°F can be used to produce ice by use of a heat exchanger submerged in a thermal storage tank containing water. The volume of the ice thermal storage tank is equal to the volume of the coil (roughly 10 percent of total volume), plus the volume of water, plus about 10 percent of the total tank volume to allow for expansion when the water turns to ice (a 9 percent volume increase). With an internal melt system, no water leaves the atmospheric (nonpressurized) thermal storage tank as it does with an external melt system. How much ice is produced is a function of the type of ice storage system. Internal melt ice-on-coil thermal storage systems are typically more efficient (volumetrically, and sometimes energywise as well) than external melt ice-on-coil thermal storage systems; but internal melt systems cannot produce the nearly constant and low discharge temperatures that can be achieved with the external melt systems.

Encapsulated Ice

Encapsulated ice systems circulate brine or chilled water-glycol through a thermal storage tank full of small containers (either spherical or rectangular) that contain water. In the charging mode, the cold circulating chilled water-glycol supply at approximately 25°F flows around the containers causing the water inside the containers to turn to ice. The containers are designed for expansion. During the discharge mode, the warm chilled water-glycol or brine return flows through the thermal storage tank around the containers of ice thereby melting the ice inside the containers and in the process producing cold CHW supply. Though physically different in details, the concept and operation of an

Figure 3-2   Packaged internal melt ice-on-coil system (courtesy of CALMAC Corporation).

encapsulated ice system has similarities to an internal melt ice-on-coil system; both systems have water that is frozen and melted without ever leaving its container; and both systems use a water-glycol or brine mixture for both charging and discharging.

With an encapsulated ice system, the thermal storage tank can be of any size, shape, and orientation, and can be either a pressurized tank or an atmospheric tank, which is less costly than a pressurized tank. Advantages of encapsulated ice thermal storage systems include the relatively larger contact area and the smaller required thermal storage tank size. Disadvantages include energy use (if the ice containers are thick) and warmer, less constant discharge temperatures compared with external melt ice-on-coil systems.

Ice Harvesters

Ice harvesters are one of the oldest active thermal storage systems and are similar to icemakers seen in hotels, restaurants, and hospitals. Most common systems produce ice on vertical plates (although some systems use vertical tubes), which then defrost repeatedly to drop ice into the thermal storage tank below. One difference of ice harvester systems compared to the ice-on-coil systems is that the ice-making heat exchanger coil is not located in the thermal storage tank. Also, unlike ice-on-coil systems, ice harvesters require separate refrigerant ice-making equipment for the thermal storage process. During the charging mode, CHW is pumped from the bottom of the thermal storage tank up to and distributed over the top of the ice making plates (or tubes). After the ice has built up to about a quarter of an inch to less than a half an inch, the ice-making equipment enters the defrost cycle (hot gas bypass) and uses hot high-pressure refrigerant (from the compressor discharge) flowing in contact with the inside of the harvester plates (or tubes) to melt the surface of the ice at the boundary of the plate (or tube) and the ice, so that the ice breaks free and falls by gravity into the thermal storage tank below. Some ice harvesting systems, instead of the defrost cycle, use mechanical means (e.g., scrapers) to remove the ice from the surface of the ice-making plates.

A separate CHW distribution pump distributes the 32°F melt water from the bottom of the ice thermal storage tank directly to the cooling loads and back to the top of the ice harvester plates, where ice is made again (partial storage mode), or the warm return CHW falls back into the thermal storage tank if the ice-making equipment is off during the on-peak period, for example. The ability to produce ice continuously is an advantage of the ice harvester systems over other ice storage systems, which must go through a discharge cycle to melt the ice. Because of the extremely large heat transfer surface of the ice pack within the tank, ice harvesters also have the ability to be discharged at a rapid rate (as fast as 30 minutes) while still maintaining relatively cold discharge temperatures , and, therefore, can be a good application for emergency cooling systems. The ice harvester thermal storage tank can be located below grade with the ice-making equipment located directly above it, at grade. Disadvantages of ice harvesters can include typical lower energy efficiency, higher complexity, and higher maintenance costs. In addition, extended periods without fully melting the ice pack can result in "agglomeration" into solid ice masses, adversely affecting subsequent water flow and heat transfer.

Ice-Glycol Slurries

Ice-glycol slurries use a mixture of glycol and water and have ice crystals in suspension in the fluid. When the mixture reaches a certain temperature point, ice crystals tend to separate and can be stored in a thermal storage tank. Ice slurry can be made in a number of different ways and has been made by spraying a liquid refrigerant and water together inside a low-pressure tank, where the water forms instant snow-like crystals as the refrigerant evaporates. The vaporized refrigerant is drawn back to a compressor. After the high-pressure gas from the compressor is condensed to a high-pressure liquid, the refrigerant is returned back to the low-pressure tank. Another method uses a refrigerant shell-and-tube heat exchanger with an agitator located above a thermal storage tank, with the ice slurry falling into the storage tank.

Ice slurry technologies have some advantages over other ice systems including the fact that there is no need to build an ice layer on the heat transfer surfaces as with ice-on-coil systems or as with ice-harvester systems, which, as discussed, reduces heat transfer and causes the refrigeration compressor to work harder and use more energy. The ice slurry can be pumped directly to cooling loads, which can reduce the volume of water needed to be pumped due to the higher delta-T.

Another direct ice slurry production method has been developed in Israel and used primarily in South Africa, Japan, and Europe, where very large, low head impellers inside a vacuum chamber compress and evacuate water vapor from the vessel, which operates near the triple point of water (32°F and a deep vacuum) to produce ice. Because the specific volume of water vapor at that deep vacuum pressure is so great, the impellers used must be very large. The ice slurry is pumped from the ice generation vessel to a storage tank, which can be near to, or far from, the ice generation equipment. The ice slurry storage tank can contain up to 50 percent ice and can be melted rapidly while maintaining a relatively constant and very low discharge temperature.

Ice systems have several major advantages over CHW storage systems. In a CHW storage system, if the CHW return comes back with a temperature lower than the design's, the lower temperature derates some of the thermal storage capacity. In an ice or other phase change material system, the phase change occurs at a specific constant temperature. That means, using ice as an example, the CHW can only be cooled to approach the 32°F phase change point. Therefore, if the designed 10° delta-T system achieves only a 2° rise in the chilled return water, there will still be only the amount of ice melted to meet the CHW return temperature. Hence, no ice is wasted (though distribution pump energy still is) by improper CHW control or building system performance. Additionally, stratification loss does not tend to be a problem with ice systems.

An additional factor favoring ice is the large amount of energy at the phase change point. As previously noted, it takes 144 Btu removed to change 1 lb of water from liquid to solid (latent heat of fusion). It would take about a 144° temperature change (sensible heat) in 1 lb of water to equal that amount of energy, and CHW sensible storage systems seldom have temperature differences of over 35°F. The actual volume differences are not as great as that would suggest, because, as noted, no ice storage system is 100 percent ice. In addition, due to different design approaches, sometimes an ice storage system requires greater footprint (ground or plan area) than a comparable CHW storage system.

Two major disadvantages of ice systems are that they tend to require more energy than conventional CHW systems and often require specialized ice production and storage equipment.

Therefore, each thermal storage technology has inherent advantages and limitations. Generalizations can be made and used as approximate rules of thumb, as presented in Table 3-3. Of course, any generalizations should be viewed with some caution, as a fuller understanding of the technologies is important to optimally select and employ

Latent heat	Sensible heat storage systems		
Characteristic	**Ice**	**Chilled water**	**Low-temp fluid**
Volume	Good	Poor	Fair
Footprint	Fair	Fair	Good
Modularity	Excellent	Poor	Good
Economy of scale	Poor	Excellent	Good
Energy efficiency	Fair	Excellent	Good
Low-temp capability	Good	Poor	Excellent
Ease of retrofit	Fair	Excellent	Good
Rapid discharge capability	Fair	Good	Good
Simplicity and reliability	Fair	Excellent	Good
Site remote from chillers	Poor	Excellent	Excellent
Dual use as fire protection	Poor	Excellent	Poor
Storage material cost	Low	Low	High

TABLE 3-3   Inherent Characteristics of Thermal Storage

thermal storage for specific applications. In particular, note that each thermal storage technology is rated as excellent in some respects, yet poor in others.

The economy-of-scale issue (in which the unit cost of CHW storage tanks drops dramatically with increasing volume, while ice storage generally uses modular equipment with relatively little economy of scale) drives most large applications to employ CHW storage and most smaller applications to employ ice storage, though there are exceptions in both cases. A survey of hundreds of cool storage applications (Potter, R. A. et al., 1995, "ASHRAE RP-766: Study of operational experience with thermal storage systems") found that there are approximately 10 times as many ice storage installations as CHW storage, yet the water storage systems are on average about 10 times as large as the ice storage systems (nearly 20,000 ton-hours versus about 2000 ton-hours); thus, each technology (ice and water) provides a roughly equal amount of storage capacity and load management in the market.

Other Phase Change Materials

Phase change materials (PCM), although not as prevalent as ice thermal storage systems, have been used for many decades. Ice, of course, is a PCM; however, the term "PCM" is typically used for one of the following four types of materials:

- Eutectic salts
- Salt hydrates
- Organic materials (paraffin waxes, fatty acids)
- High temperature salts.

While liquid-to-gas and gas-to-liquid are phase changes, thermal storage PCM typically deal with solid-to-liquid and liquid-to-solid changes. The change from a liquid to a vapor involves a large change in volume, whereas liquid-solid phase changes involve much less volume change. PCMs are used for both heating thermal storage, particularly with solar heating, and for cooling thermal storage. Melting or freezing points of PCMs vary from −100°C to more than 1000°C depending on the type of PCM, with commercial systems presently available for more than 100°C. Ideal PCMs have a high heat of fusion, a high density, are nontoxic, chemically stable over extensive thermal cycling, and melt over a narrow temperature range (some PCMs do not change phase at a specific temperature).

Glauber's salt (sodium sulfate or Na_2SO_4) is a eutectic salt, which encapsulated in individual containers similar to encapsulated ice, has been used in thermal storage systems. A common thermal storage eutectic salt melts at 47°F, and standard chillers producing 40 to 42°F can charge the eutectic thermal storage system. Eutectic salts have a heat of fusion of about 40 Btu/lb.

Salt hydrates have storage densities between 3000 to 4000 Btu/ft^3 of storage at the phase change point. Some salt hydrates have proven to be chemically unstable.

The melting point for organic waxes varies considerably depending upon the length of the hydrocarbon chain. Organic materials often have less attractive thermal properties and high material costs compared to other PCMs.

Seasonal Thermal Storage Systems

Seasonal thermal storage systems, including the use of aquifer thermal energy storage (ATES), have been designed and constructed in the Pacific Northwest, Canada, Sweden, and the People's Republic of China; however, seasonal thermal storage works best where there are two distinct seasons, a summer requiring cooling, and a winter requiring heating, with minimal spring and fall periods, and in high latitude locations where solar energy is available for long periods of the day during the summer. Seasonal thermal storage can either be a cooling system, a heating system, or a hybrid heating and cooling system. Water is often used as the thermal storage media and can be pumped to or from underground insulated storage tanks, caverns, or naturally occurring underground aquifers.

Underground earth/rock strata can also be used for long-term or seasonal thermal storage. For cooling, as an example, a source well and a recharge well combined with a cooling tower might cool water during the winter and, over a long period, reduce the temperature of a confined underground gravel strata or contained aquifer. During the summer, circulation of water through the strata might exchange energy to provide CHW. There are many variables with this system, and much research is needed to develop optimum applications. Several projects have used this approach and one of the oldest in operation was installed in Portland, Oregon, in 1947.

CHAPTER 4

Sensible Thermal Storage Systems

Introduction

Chapter 3 discusses both sensible and latent thermal storage and presents some of the basics of many types of thermal storage systems. In addition, as discussed, adding energy to a material causes a rise in its temperature (sensible) or a change in state (latent). The principle that any material can either change temperature or change state, in one sense, makes many materials possible candidates for thermal storage taking into consideration the items listed and discussed in Chap. 1 regarding thermal storage media. This chapter addresses some actual applications using sensible thermal storage.

In sensible thermal storage, such application is limited to liquids and solids. Sensible storage is used for both heating and cooling operations, and for solar thermal storage operations. In a heating application, some process is used to add energy to the storage material. Most heating storage applications require that the storage material be heated hotter than the application served by the thermal storage system. Heat is transferred to the application by conduction, radiation, or convection. Common applications include space heating, domestic hot water production, swimming pool heating, and process heating.

Cooling sensible thermal storage on the other hand requires removal of energy from the storage material, which cools the material. Similar to heating thermal storage, most cooling thermal systems require the thermal storage material to be cooler than the application served by the thermal storage system. Heat is transferred to produce a cooling effect by conduction, radiation, or convection. Cooling applications are used primarily for space cooling and dehumidification and for process cooling and for combustion turbine inlet air cooling.

With either heating or cooling thermal storage, there is usually some intermediate material such as water, water-glycol or water-brine, or air to transport or conduct the heat transfer.

Sensible thermal storage systems offer some advantage over latent thermal storage systems because sensible thermal storage systems operate over a range of temperatures, whereas latent thermal storage systems typically provide most of their storage at a single temperature, the phase-change temperature. To produce ice, as an example, requires production of all the storage at or below 32°F (0°C). The sensible thermal storage advantage to operate over a range of temperatures can be maximized, for example, by using multiple chillers in series such as a series-counter-flow arrangement. With series chiller arrangements, the lead chiller(s) need not cool the chilled water to the final temperature, so less total energy is needed than is required to produce the final storage temperature. The use of series chillers is particularly an advantage when a sensible storage system operates over a large temperature difference (delta-T). In addition, sensible cool storage can operate at higher temperatures than ice systems, which, all other things being equal, requires less cooling production energy.

Factors Affecting the Selection of Sensible Thermal Storage Materials

Although almost any material could be used for sensible thermal storage, as outlined in Chap. 1, there are a number of factors that favor certain materials over others for use as a thermal storage medium. The combination of these factors has resulted in certain materials being commonly used in sensible thermal storage. Solids typically used for sensible thermal storage are building mass, gravel or stone, ceramic bricks, or the earth itself. Liquids used for sensible thermal storage are usually water or water plus some additive(s) to change the maximum density point and freezing point compared to those of pure water, and oil or molten salt.

Solid Sensible Thermal Storage Systems

Solid sensible thermal storage systems use a change in a material's temperature to store thermal energy (or technically in the case of cooling thermal energy storage to store an energy sink). There are three typical types of solid thermal storage systems:

- Use of building mass
- Use of the ground
- Use of a specially constructed system with container, storage material, and a circulating system

Use of Building Mass for Thermal Storage

Building mass has been used for thermal storage at least since the ancient Romans. Roman baths, as an example, were constructed with raised floors and pools with an open circulation space below the floor. Fires heated air, and the air and smoke were drawn through the space below the floor, heating the floor and the water above. Because the floor, structure, and water had mass, the heated floor and pool would contain the heat after the fire was out. However, some thermal mass uses are even older.

It appears that the caveman learned that if a cave has a south-facing opening, the low winter sun angles would heat the back of the cave. The mass of the cave wall would heat the space, somewhat, even after the sun went down. The thermal mass principle (mass stores heat and keeps cool for a time), when learned, was later specifically incorporated into building design. The Brookhaven House described in *Sun, Wind and Light*[1] is an example of using a high-mass interior wall and controlled external glazing to both heat the home with solar energy and store heat in the wall for heating when the solar gains are not available.

Building mass is commonly used for thermal storage without even considering that thermal storage is being done or why the process works. Some buildings are ventilated with outside air at night to cool the building mass. The colder building materials are able to absorb some of the daytime heat gains, reducing the amount of mechanical cooling needed during the day.

Many buildings with large fluctuations in load use the building mass to help meet peak cooling loads. Large restaurants and meeting halls, and even sports arenas, often "precool" the space to reduce the peak cooling load. The spaces start out somewhat cooler than optimum and often finish somewhat warmer than optimum, but much less cooling capacity is needed and electrical demand is reduced.

Building mass has also been used for sensible thermal storage in other ways. The Crop Science Building at Oregon State University was designed with hollow core precast floor panels, so that cold night air can be forced through hollow tube cores in the floor at night to precool the structure, when cooling is expected to be required the next day. When heating is expected to be needed in the near future, room air can also be returned or exhausted through the hollow core structural mass, which warms the mass and can offset heat needs. Figure 4-1 shows a schematic of the Crop Science system. Active building mass as a thermal storage element is well established in Europe.

However, excessive use of building mass to offset cooling and/or heating needs can be overdone and produce occupant thermal discomfort. Many, perhaps, have gone into a building in the winter when the heating systems have been off over the weekend. The cold building mass makes the space uncomfortable. Indeed, even when the air is up to

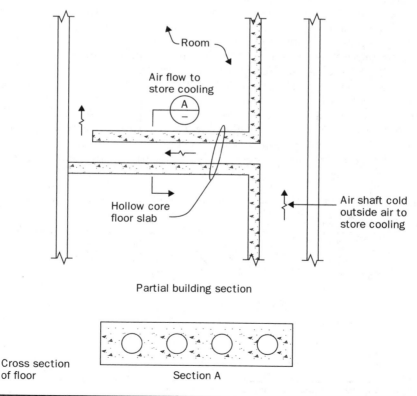

FIGURE 4-1   Active building mass for thermal storage in the Oregon State Crop Sciences building.

the desired temperature, body thermal radiation to the cold structure can still cause thermal discomfort.

Use of the Ground for Sensible Thermal Storage

The ground has been used for thermal storage especially for longer term (such as annual) thermal storage. The goal of such thermal storage systems is usually to heat the ground with "free" or recovered heat including space heat rejected during the warmer part of the year. The stored heat is then recovered and used during the colder part of the year. This concept can reduce the winter heat needed and at the same time reduce the summer mechanical cooling needed. Thermal storage systems have also been designed to cool a transfer medium such as water with cold winter air and circulate it in the ground to cool the ground. The systems then reverse in the summer to draw on the stored cooling from the ground, and the heat from the building is stored in the ground. Most often, this concept requires two or more wells and reversing operation depending on whether the system is being used for heating or cooling. Figure 4-2 shows a simplified schematic of a ground mass thermal storage application.

To be successful, a ground mass thermal storage system cannot be a great deal warmer or colder than the normal earth temperature, as the greater the temperature differential, the faster or greater the heat loss. Heat is conducted to the earth, and large temperature changes would always lose or gain heat to or from the surroundings. Some systems, when used with moist soil cause freezing around the heat exchange piping in the ground, which increases the amount of cooling storage that can be obtained, though such a system is more properly considered to be a latent heat thermal storage system. Because ground temperatures are usually near the average soil temperature, often a heat pump is needed to store heat in or obtain heat from the soil.

Ground source (earth-coupled) heat pumps provide at least some limited thermal storage, when the systems provide both space heating and space cooling. Many ground source heat-pump systems incorporate heat exchange piping in either trenches or vertical bore holes similar to wells. With this type of ground source heat-pump system, a circulating medium such as a glycol-water mixture is circulated in the piping system. When the heat-pump system is heating the spaces served, energy is taken from the circulating fluid by the heat pump, reducing the temperature of the circulating fluid. The

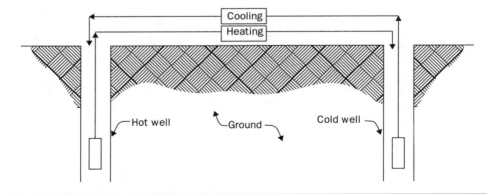

Figure 4-2 Ground mass thermal storage system schematic.

cold circulating fluid from the spaces served is then warmed by contact with the earth, when the medium is colder than the earth. When the heat pump is cooling the space served, the opposite occurs, with the energy taken from the space by the heat pump added to the circulation fluid. When the circulating fluid is warmer than the surrounding ground, energy is transferred to the ground, cooling the fluid and warming the ground.

An ideal ground source heat pump application would have an exact balance over time between cooling and heating loads; however, this ideal balance seldom occurs, and the systems often must make up heat or lose additional heat to the earth itself. Sometimes, additional boiler or solar-produced heat is needed, and, in other cases, cooling towers have been needed to remove excess heat from the system.

A factor that affects thermal storage and the application of the ground source heat-pump system is how well heat is conducted into and recovered from the earth surrounding the heat exchange piping. For energy to flow (i.e., heat to transfer), there must be a temperature difference between the ground and the circulating fluid in the pipes. Over time, the temperature difference needed to transfer heat can increase as energy must be conducted over a greater distance within the earth. Changing the temperature difference/delta-T of a sensible thermal storage system directly affects the thermal storage capacity, because, as shown in Chap. 3, the thermal storage capacity is directly proportional to the temperature difference.

Reversing the ground source heat-pump application so that sometimes the system is heating and sometimes the system is cooling improves the operation of the ground source heat-pump system. With a reversible ground source heat-pump system, when the heat-pump system rejects heat to the ground, there is thermal storage in the ground. When the heat-pump system requires heat and thermal energy is withdrawn from the ground by the system, the stored heat is recovered. Figure 4-3 shows a ground source heat-pump system using earth as a thermal storage system.

One variation of the ground source heat pump is a system that uses multiple wells that are reversed annually. This multiple-well system uses well water to transfer energy between the ground (gravel layers) and the building systems. A clinic in Eugene, Oregon, is an example of a multiple-well heat-pump thermal storage system and has a

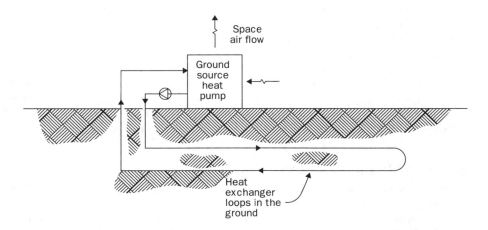

FIGURE 4-3 Ground-coupled heat-pump system using earth as a thermal storage system.

two-well system. The clinic is served by many small water-to-air heat pumps connected to a common heat loop pipe. The heat loop piping connecting the heat pumps is also connected to a plate-and-frame heat exchanger with well water on the other side of the heat exchanger. When the building heat loop requires more heat, well water is circulated from the "warm" water well to the heat exchanger and returned to the "cold" water well. The heat obtained is drawn from the gravels and heat of the earth itself.

In the above example, the need for heat is essentially limited to the winter period, and, during that time, the discharge of cold water into the cold recharge well cools the gravels and earth in the area of the well. When the heat balance of the building requires a net heat removal (i.e., overall cooling), the wells are reversed. Water from the cold well passing through the plate-and-frame exchanger cools the building circulation medium and raises the temperature of the well water. The warmed well water is then returned to the warm water well. All of the well water is maintained in the ground at the site and it serves only as a medium to take heat out of or reject heat into the gravel below the site.

Upon start-up, the gravel below the building provides and absorbs the heating and cooling needs of the building, respectively. Over time, this operation results in different temperatures at the two wells because of thermal storage. Having warmer well water available in the winter, when the building needs heat, reduces the purchased electrical energy needed by the heat pumps in heating mode. Likewise, having colder water available in the summer, when heat is being rejected by the building, reduces the purchased electric energy needed by the heat pumps in cooling mode. The earth provides any annual imbalance between heat needs and heat rejection needs.

Gravel as a Sensible Thermal Storage

Gravel, either in a constructed enclosure or in the ground as a functional structural material, is used for thermal storage. Many systems have been designed to use this type of thermal storage. Most of them are either a part of a solar thermal system, or a so-called free-cooling system. With solar-gravel thermal systems, air heated by the sun is forced by a fan or by natural convection through a large bed of gravel, which heats the gravel or rock storage during the sunlight period. At nights or during cloudy weather, the airflow is reversed and heat is delivered to the space or building that the thermal storage system was designed to serve. Gravel is also used to store cooling. Cold night air is drawn through the gravel storage to cool the material. Airflow is reversed, when cooling is needed by the building served.

Although the above-described type of gravel bed system usually uses outside air directly, it is possible to provide greater thermal storage by mechanically cooling the air. Mechanically cooling the air provides more thermal storage for a given size gravel bed, whereas the mechanical cooling uses less expensive nighttime or off-peak electrical power. Figure 4-4 shows a gravel thermal storage system.

Liquid Sensible Thermal Storage Systems

The above-described sensible thermal storage systems use a solid for sensible thermal storage and a liquid as a heat exchange/transfer media. More commonly, liquids are used for sensible thermal storage. Although there are advantages and disadvantages for each type of medium for use in sensible thermal storage, there are more advantages to the use of a liquid than advantages to use of a solid for sensible thermal storage.

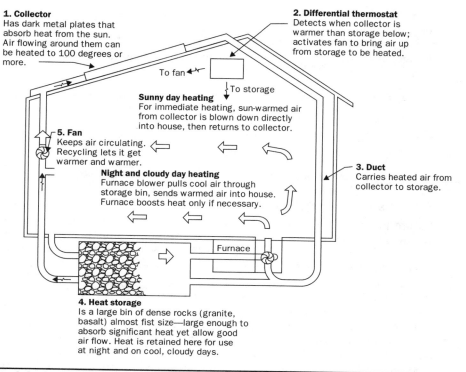

1. Collector
Has dark metal plates that absorb heat from the sun. Air flowing around them can be heated to 100 degrees or more.

2. Differential thermostat
Detects when collector is warmer than storage below; activates fan to bring air up from storage to be heated.

To fan

To storage

Sunny day heating
For immediate heating, sun-warmed air from collector is blown down directly into house, then returns to collector.

5. Fan
Keeps air circulating. Recycling lets it get warmer and warmer.

Night and cloudy day heating
Furnace blower pulls cool air through storage bin, sends warmed air into house. Furnace boosts heat only if necessary.

3. Duct
Carries heated air from collector to storage.

Furnace

4. Heat storage
Is a large bin of dense rocks (granite, basalt) almost fist size—large enough to absorb significant heat yet allow good air flow. Heat is retained here for use at night and on cool, cloudy days.

FIGURE 4-4 Gravel thermal storage system.

Some of the factors affecting the advantages and disadvantages of using a liquid versus solid include how energy is moved into and out of the thermal storage medium.

With a solid, energy is typically applied or removed from the surface of the material and heat must be conducted into or out of the storage medium. Only after a period of time does the interior of the solid thermal storage medium approach the temperature of the outer surface. The amount of time and how great the challenge is to bringing the interior of the solid to the desired temperature depends on the size of the individual storage elements and the thermal conductivity of the thermal storage material and the temperature difference. Regardless of the size and thermal conductivity of a material, at the beginning of either a storage or recovery cycle, the rate of heat transfer is the greatest, and the heat transfer rate degrades over time as the storage medium approaches its final storage temperature throughout the thermal medium.

A liquid, of course, does not have a fixed shape, and so a liquid is not subject to the heat transfer challenge just described. Additionally, most solid storage systems require an intermediate transfer fluid to transfer energy between the energy production system (e.g., chiller, boiler) and the end use (e.g., building heating or cooling coils). With a liquid thermal storage system, the same fluid that provides the thermal storage is often used to transport the energy to thermal customers or users (i.e., the end use). In other cases, there may be a plate-and-frame heat exchanger or some similar device to isolate the thermal storage from the energy transport system.

One of the disadvantages of a liquid storage system is the need to maintain a separation between the colder fluid (supply water in cool storage, return water in heat storage) and the warmer fluid (return water in cool storage, supply water in heat storage). As discussed in Chap. 3, there are several ways the challenge of temperature separation has been addressed including:

- Naturally stratified thermal storage systems (lakes, ponds, and ocean)
- Series or labyrinth tanks
- Multiple tanks employing the "empty tank" method
- Membrane or diaphragm tanks
- Stratified thermal storage tanks (by far the dominant approach used today)

Natural Storage Containers

Nature provides the container for some liquid thermal storage systems. One such system uses a lake for thermal storage, where winter weather is very cold. Nature cools the lake water and the cold surface causes the water to reach its maximum density point (about 39°F) and sink to the bottom of the lake. In the summer, when cooling is required by a nearby university campus, cold lake water is circulated from the bottom of the lake, through plate-and-frame heat exchangers and returned to the warmer top of the lake. The large mass and depth of the lake is essential to providing annual thermal storage in this way, as the surface of the lake naturally warms during the spring, summer, and fall period, whereas the deep lake water remains cold year round. This lake cooling system is able to provide campus cooling with pumping energy only, which is approximately one order of magnitude less than required for mechanical cooling. Similar thermal storage systems have been designed even in Hawaii using cold ocean water taken from deep offshore, whereas others have been in operation in Canada and Scandinavia using deep harbor seawater. Ponds have also been either constructed or natural ones used for both heating and cooling thermal storage. Experimental ponds using layered salt solutions have been constructed in Israel near the Dead Sea and used to collect and store solar energy for use in heating.

Series or Labyrinth Tank Thermal Storage System

Series thermal storage tank designs are multiple tanks (or multiple sections within a single tank) arranged with the bottom of one tank connected to the top of another tank. This piping arrangement confines any mixing to a single tank and creates all the thermal stratification advantages of a single tank as tall as the sum of all the tanks. Typical connections can be a simple baffle and weir arrangement, which turns one large tank into several smaller tanks in series. Figure 4-5 shows the arrangement of a series tank installation. Flow direction is reversed between charging and discharging cycles.

Series tanks often are constructed using prefabricated materials such as fiberglass storage tanks of a type used for fuel storage. Such tanks are often installed below parking lots. Although quite popular at one time, the series tank concept is seldom used today except on very small thermal storage systems.

Labyrinth Tank

A labyrinth tank solution is a variation of the series thermal storage tank solution. It is a design of many tanks (or tank sections) arranged in a way that water from one tank

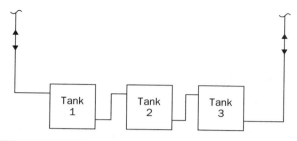

Figure 4-5   Series tank thermal storage system.

flows to the next and is similar to the series tanks described above. The two primary differences between series tanks and the labyrinth solution are the number of tanks and the locations of connections. Although series tanks are typically connected bottom to top, labyrinth tanks are typically bottom-to-bottom and top-to-top connected. Water flowing in through the tanks flows from the bottom of the first to the bottom of the second and from the top of the second to the top of the third and from the bottom of the third to the bottom of the fourth and so on through many tanks. This arrangement does not depend in any way on density difference between the supply and return water, but rather depends on physical separation of the return water tank, the mixed water interface, and the water supply tank. This design would reduce the mixing or interface area to a very small amount. If water flows from the bottom of one tank to the top of the next, the effect is similar to a very tall tank. Figure 4-6 shows a labyrinth tank thermal storage system. Again, flow is reversed between charging and discharging. Labyrinth tanks were primarily employed in Japan within the subbasement foundations of tall buildings, for which an "egg crate" style of structural design provided a ready-made means for adopting the labyrinth storage.

Multiple Tank "Empty Tank" System

The empty tank concept is perhaps the easiest to understand, and is also easier to explain the operational needs, which occur in every type of water thermal storage system. As shown in Fig. 4-7, the empty tank concept employs multiple tanks or sections of a single tank (two or more), including one more tank (or tank section) than the total liquid storage volume. For example, if there are 10 tanks holding 10,000 gallons each (total volume 100,000 gallons), an extra 10,000 gallon tank is provided. Valves on each tank direct which tank the liquid (chilled water usually) will be drawn out of and which one tank will serve for the return water. After an "empty" tank has been filled, another tank has become empty; the valves are adjusted to now draw water from a previously unused full tank and return the water to the new "empty" tank.

Each of the tanks must have the same volume, and, as noted, the system total water volume is less than the total tank capacity, by the capacity of one tank. In a cooling example, the current "empty" tank and the current active "full" (supply) tank act as the interface between chilled water return and chilled water supply. The empty tank design has a number of valves that are used to control from which thermal storage tank water is withdrawn and to which it is returned. Return chilled water from the building is piped to one of thermal storage tanks so that water not returning to the chillers, returns

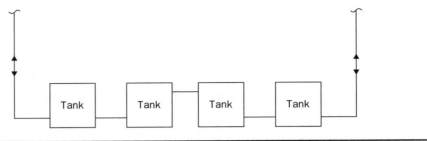

FIGURE 4-6 Labyrinth tank thermal storage system.

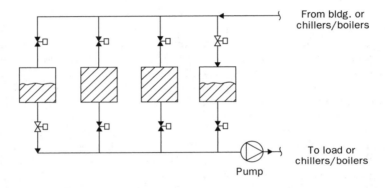

FIGURE 4-7 Empty tank thermal storage system.

to the thermal storage tanks. Chilled water produced by the chillers is piped in an arrangement that any water not used by the buildings flows to the supply thermal storage tank. Valves are able to withdraw water from any tank either to provide building cooling or to be cooled by the chiller.

Continuing with the cooling example, in operation with the thermal storage tanks fully charged with chilled water, an empty tank is available in the tank line. Any return chilled water not flowing to the chiller would return into that empty tank. The valve in another tank would be open to the chilled water supply system so that if the facility cooling demand were to exceed the instantaneous chiller production, the additional cooling need could be met by chilled water withdrawn from that tank.

During high cooling load periods, water is withdrawn from the storage tank. As the storage tank is depleted, its supply valve is shut and the supply valve to a full tank is opened and the return valve to the tank just emptied is also opened. This means water is supplied from the full tank and returned to the tank just emptied. In use, the storage tanks repeat this operation moving from tank to tank. When all the tanks have all been used, the tanks contain chilled water return except of course for one empty tank, and the thermal storage system is fully discharged. To use the system again, the water in the tanks must first be rechilled and thermal storage system charged.

To charge the empty tank thermal storage system, the operation reverses. If the chilled water return from the buildings is less than the chiller production, the additional chilled water return to the chiller plant will come from the thermal storage tanks. The additional chilled return water flowing to the chiller is provided from the tanks in a reverse operation to the withdrawal for building use.

The heating system concept is analogous.

The empty tank concept makes it easier to understand what happens in any chilled (or hot) water storage system and the need for variable chilled water flow. The building chilled water flow must vary in proportion to the cooling load to achieve the proper relation between storage and withdrawal. Higher flows in the system than are produced by the chiller (or heat production device, as the case may be) result in the use of some tank water. Use of less water than produced by the chiller, results in storage of the surplus capacity.

The concept also makes it easy to see the effect of less than full temperature rise of the chilled water on the system storage capacity. Water that has only a small temperature rise takes the same volume as water with a full temperature rise, but the storage amount (in potential Btu of cooling) is far less.

The empty tank concept eliminates mixing of supply and return water, but it does require extra storage volume, higher unit capital cost (associated with smaller tanks), higher complexity of piping and valves, more pumping horsepower (other factors being equal), and it introduces air into the system as water drops into the top of an empty tank, which can increase corrosion and microbiological activity (problems have occurred in the past). The cost, performance, and complications are great enough that few empty tank thermal storage systems have been designed, although one such large empty tank system was designed for Arizona State University in the mid-1980s using six 1.1 million gallon tank sections to contain 5.5 million gallons of chilled water.

As noted, empty tank systems may be used for both heating and cooling thermal storage, however, cooling is the most common application. One example of an empty tank heating thermal storage is an underground insulated storage tank for the States Street wading pool at the Eugene, Oregon Parks and Recreation District. Water is heated by solar collectors and stored in the underground storage tank. When desired for wading, the water is pumped out into a wading pool. After use, it is returned to the tank. Another example is a residence with an in-the-bedroom hot tub. An over-side hot water heater-storage tank heats and stores the large volume of hot water needed. When one wants to use the hot tub, the water is pumped into the tank. After use, the water is returned to the tank. Because the period of use is small, this saves most of the heat losses of a normal hot tub, reduces the heater size needed, and eliminates most of the moist (water vapor) losses to the bedroom, which would cause condensation, mold, and dry rot in the wood construction of the bedroom.

Membrane or Diaphragm Thermal Storage Systems

Membrane thermal storage systems are a patented concept used extensively in Canada in the 1980s. Used for cooling storage, membrane separation systems consist of a single tank with a flexible membrane (or occasionally a rigid diaphragm) to physically separate the chilled water return (above) and chilled water supply (below). The diaphragm or membrane moves up and down as the tank is charged and discharged. When the membrane system works properly, it eliminates any possible mixing of supply and return water as the membrane moves with the changes of relative volume in the cold and warm water. Although the membrane eliminates direct mixing, heat can still be conducted through the barrier, and, over time, there is an energy interchange, which produces an effect similar to mixing.

The membrane design concept was offered as a solution to the interface problem. It does not, however, eliminate the complete problem. In part, this solution does not solve the interface problem because, as noted, it does not eliminate thermal exchange between the hot and cold side, as the membranes are typically not insulated. Additionally, some space is needed to introduce and to withdraw water, which initially is the interface zone, and insertion of a membrane cannot eliminate that space. Also, the solution is somewhat mechanical and is subject to failure (e.g., tearing or structural failure of the membrane or diaphragm, and the membrane blocking pump suction lines causing pump cavitation). The University of New Mexico conducted field tests of flexible membrane systems, showing they do not substantially increase the performance over a good thermal stratification design without a membrane.

Stratified Thermal Storage Systems

By far, the most common sensible thermal storage systems today are stratified thermal storage systems. The popularity of vertical stratified thermal storage systems is, in part, due to the simplicity of the system. Moreover, although the basic stratified thermal storage system is simple in concept and construction, the application of the system to real-life situations requires a great deal of care in planning and design. As a consequence, stratified thermal storage is most often applied to larger systems with individually engineered designs.

Stratified thermal storage depends on the density difference between colder and warmer water. This temperature stratification has been used for centuries in hot water storage but, more recently, has formed the preferred solution in chilled water storage. The big difference between hot water thermal storage and cold water thermal storage is the relative power of the density differential to naturally separate supply from return water. In hot water or heating water systems, usually, there is a very large temperature difference between the colder water and the warmer water. That large temperature difference results in a large difference in density, which provides substantial buoyancy for the warmer water to float on top of the colder water.

In contrast, with CHW thermal storage system, as water gets colder, it approaches its maximum density point (at about 39.4°F or 4.1°C). The closer the water gets to the maximum density point, the less is the difference in density for each degree of temperature change. Although this complicates things somewhat for stratified cooling water thermal storage, designers have found that, with careful tank and flow diffuser design, most of the tendency of chilled supply and return water to mix can be reduced. The simplicity of stratified storage designs has led to the concept being developed as the preferred solution.

One modification to the above comment regarding the maximum density point of water, as discussed further in this chapter, is that when certain chemicals are mixed with the water, both the freezing point and maximum density point of water can be lowered, creating a low-temperature fluid (LTF), which can be stored in stratified storage at temperatures well below 39°F. Although adding chemicals increases the cost of the storage medium (and low-temperature operation increases chiller energy use), the advantages of a colder supply temperature, a larger delta-T, and a smaller volume thermal storage tank can outweigh the disadvantages, in some cases. Figure 4-8 shows

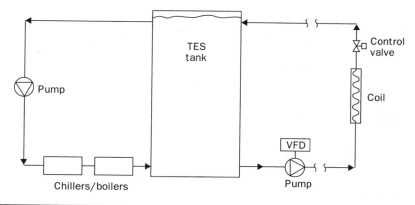

FIGURE 4-8   Stratified thermal storage system.

a simplified stratified storage system. This simplified concept can serve for either heating or cooling thermal storage except that supply and return diffusers are reversed.

Sensible Water Thermal Storage System Needs

A sensible water thermal storage system specifically needs a number of elements to function properly to be effective as discussed in the following paragraphs.

First, the system capacity and flow through chillers or boilers must provide the design supply water supply temperature to the thermal storage system. The thermal storage charge rate is equal to the rate of supply production less the capacity used instantaneously. On the other hand, if the load exceeds the instantaneous chiller or boiler production, the difference is supplied by the thermal storage system. The entire thermal load can sometimes be met by thermal storage with no boiler or chiller capacity produced at that time of use. If there is no thermal load being served, all the boiler or chiller capacity produced goes to thermal storage.

Second, a hydronic system should only use the limited volume of water required to meet the heating or cooling load. For older systems, chilled water or boiler water were often piped via constant flow pumping to coils served by three-way valves, which will not work well with sensible water thermal storage. With sensible water thermal storage, modulating two-way control valves are required to use only the amount of heating or cooling water needed to provide the necessary thermal capacity. The two-way valves must be able to close against full pump shut-off head and modulate properly without allowing any excess flow. The use of modulating two-way valves in turn typically requires that pumps serving the load are variable-volume.

Third, a sensible thermal storage system requires a facility or building that obtains return water temperatures near the design temperature. When the temperature difference (i.e., the delta-T) achieved in coils or other load is less than design, the capacity of the sensible thermal storage system is reduced accordingly.

Fourth, a system must be able to separate used return water from the unused supply water, which is achieved by one of the methods previously described.

Fifth, a system should develop a relatively high water temperature difference between supply and return (delta-T) for optimum savings. The delta-T affects the size and cost of the thermal storage tank needed, the piping and pump sizes required, and the pumping energy consumed.

The above issues are discussed further in the next section.

Chilled Water Thermal Storage Systems

Most of the sensible thermal storage systems designed today are cooling thermal storage systems using either chilled water or chilled water with an LTF additive. The most common cooling sensible thermal storage systems today are stratified thermal storage systems that use the density difference to maintain separation between the chilled water return and chilled water supply.

Vertical Thermal Stratification

As discussed above, vertical thermal stratification system is a method to separate chilled water supply from chilled water return and depends on the difference in density between warm chilled water return and colder chilled water supply. This density difference, although small, is enough to assure that if water is carefully introduced into a tank with warmer, more buoyant water on top, it will remain in the same relationship to the colder more dense water below it. This condition is similar to the thermal inversions that cause the winter fog within trapped low-lying areas, or to the natural thermal stratification of sun-warmed water near the surface of an undisturbed (but otherwise cold) lake. Water must be moved into and through a tank at very low velocities in order not to disturb the relative position of warmer water on top and colder water below.

Although a very good design produces only limited mixing (mass transfer), there is still some mixing that occurs on any TES tank when warm return water is first returned to a tank fully charged with cold supply water. Even if it were possible to develop a perfect interface of warm return water in contact with cold supply water, there would soon develop a transition zone as heat was conducted from the warm return water to the colder supply water (heat transfer). Heat travels from hot to cold without respect to the density change or the vertical direction. Finally, as the tank contents are cycled between cold and warm conditions, so is the tank wall structure; this creates some heat transfer between the fluid and the tank walls, primarily at the zone of temperature transition. These three factors (mass transfer within the fluid, heat transfer within the fluid, and heat transfer between the fluid and the tank walls) combine to create a zone in the storage tank where there is a temperature change from supply water temperature to return water temperature. In the case of a stratified storage system, this zone is called the thermocline. This thermocline is taken into account in thermal storage tank design in a similar way to the extra tank necessary in an empty tank design concept. Both increase the storage tank volume needed to meet the cooling load.

Some designers have tended to talk of the lost capacity due to the thermocline in terms of a percent of storage capacity, such as 10 percent. Actually, the amount of volume lost depends on many factors such as tank design, water depth, diffuser design, number of tanks, and the time the water is in storage. In most cases, the lost capacity is more related to vertical height than the percent of storage capacity. A very good design may contain the entire thermocline in 2 feet (ft) of vertical height, whereas a poor design may have as much as 8 to 10 ft. or more. Two feet of height in a 50-ft.-high tank is only 4 percent of the volume, whereas the same 2 ft. represents a 20 percent loss in a 10-ft.-high tank. This geometrical relationship encourages specifying as tall a vertical dimension in chilled water storage as can be economically, aesthetically, or practically designed (or selection of an alternative design, which achieves the same effect).

Various designs have been used to introduce and withdraw water from vertical storage tanks. Tests by the University of New Mexico for the Electric Power Research Institute and actual design experience have shown that distributed nozzle (e.g., slotted or perforated piping diffuser) designs provide the least mixing losses and the best performance. Radial flow (plate-type) diffuser designs have also been widely used with good success.

Factors That Affect Stratification

As discussed, achieving the design delta-T is an important issue to a successful thermal storage system. Less than design, delta-T will reduce the storage capacity and may also reduce the capacity of the chiller plant (with or without a thermal storage system) when less than design return chilled water is cooled to the design chilled water supply temperature. Various factors can cause this failure, such as coil performance, problems with control valves, high differential pressure near a chiller plant causing unwanted flow through forced open control valves. In addition, in some cases, with large systems, changes in flow in one building can affect the flow in other buildings and impact performance and/or chilled water temperature rise.

Unlike some other thermal storage systems, chilled water storage systems have a number of important requirements as discussed on the following pages.

Constant Chilled Water Rise

Conventional nonstorage chiller plants have often been designed to provide a certain chilled water supply temperature and accept any chilled water return temperature that the building chilled water system provides. Because older chiller plant designs required a constant flow through the chillers, chilled water distribution piping systems were usually designed to bypass any water not needed by the coils back to the chillers. That chilled water bypassing results in a fairly constant, high flow through the distribution piping systems. When cooling loads are light, most of the water is bypassed back to the chiller and the chilled water delta-T at the chiller plant is low. A chilled water storage system utilizes thermal storage capacity by raising the temperature of the stored water, which means that chilled water storage systems will not work well with systems that have three-way control valves that bypass unused low-temperature chilled water back to the return line and thus back to the thermal storage tank. It also means that chilled water control valves must close when air-handling units are turned off. Failure of the control valves to close during periods without airflow results in the flow of unused chilled water from the supply to the return and dilution of the cooling potential of the remaining tank storage charge.

Separation of Chilled Water Supply and Return

As previously noted, a separation is needed between the used chilled water return and the unused chilled water supply. This separation assures that the maximum volume of stored chilled water at the design supply temperature can be properly supplied as needed to the coils. This physical separation is not required in phase-change thermal storage solutions. Temperature separation is needed with chilled water thermal storage systems, because without temperature separation, used chilled water return mixes with chilled water supply. If any heat has been added to the chilled water return, this mixing raises the chilled water supply temperature, which can result in deterioration of the cooling coil performance and a loss of cooling ability. Very limited thermal storage can be provided by a system that does not maintain a constant low chilled water supply temperature at or near the design chilled water supply temperature.

This is very different than in a phase-change (ice or eutectic) storage solution. In the case of phase-change materials, the melting occurs in direct proportion to the load and a partial chilled water temperature rise (less than design delta-T) simply melts less phase-change material than a full chilled water temperature rise. On a chilled water thermal storage system, if the same low-temperature rise occurs in the chilled water return, the storage capacity (which is proportional to the delta-T) is proportionally reduced.

Variable-Flow Chilled Water Use

Because a full design rise in chilled water temperature is necessary to utilize the full storage capacity with chilled water thermal storage systems, a method of reducing the flow from storage at less than full-load conditions is necessary. The two ways to accomplish a full temperature design rise are either to vary the chilled water flow in proportion to the load while maintaining a constant temperature rise with variable load or to raise the supply temperature to the coils to maintain a constant return water temperature. Both concepts have been used in successful thermal storage system designs. A variable-supply temperature system mixes coil return water with building supply water at the coil to maintain a constant return water temperature from the building. At light loads, the supply temperature at the coils may approach the return temperature. Because the loads on all coils are not the same, one or more coils may require a large amount of cooling when the chilled water systems as a whole are serving a light load. This can result in inadequate cooling of some areas that have relatively high loads during light load conditions at other coils. Proper humidity control is also a concern with higher coil entering chilled water temperature. Because of these issues, and because of the reduced energy consumption possible with variable flow, variable flow is generally the better response to the challenge.

A variable-flow distribution system has modulating two-way control valves on the coils and a variable-flow chilled water pumping system that maintains a design pressure differential across the coils or maintains at least one control valve nearly fully open. Large hydronic distribution systems may require pressure-compensating control valves (also known as pressure-independent flow control valves) able to close off against high pressure to keep coil control valves near the pump from being forced open at time of high pressure differential in the piping mains.

Maintaining Equivalent or Higher Coil Capacity

Cooling coil capacity depends on many factors including entering air dry-bulb and wet-bulb temperatures, entering water temperature, quantity of airflow, quantity of water flow, and the coil design (i.e., coil area, fins-per-inch, etc.). Often it is desirable to provide lower supply water temperatures to coils when thermal storage is included than on conventional nonstorage designs. Lower chilled water supply temperatures can be combined with higher return water temperatures to produce the same coil performance as occurs with lower temperature difference in a conventional nonstorage system. A problem sometimes develops under low flow conditions. A still water film next to the tube wall can reduce the heat exchange rate, when the chilled water flow is reduced. One way to overcome this phenomenon is to recirculate a portion of the water leaving the coil back through the coil, thus maintaining high velocity for good heat exchange. That increases the supply temperature, which can reduce the coil capacity. However, if the coil does not achieve the cooling needed, the coil valve opens; therefore, the process is more or less self-correcting. With this arrangement, the chilled water return from the coil is more or less constant and the supply water temperature at the coil is adjusted by recirculating return water.

In some cases, chilled water return temperatures can also be increased with coil designs. While the typical thought is that it would take more rows and greater air pressure drop to achieve higher chilled water return temperature, that is not necessarily the case. The amount of temperature rise in a coil is also a function of circuiting the coil. Also, coil face area can be increased at modest cost, which will lower the coil air pressure drop and resultant fan energy penalty. Higher temperature rise across the coil typically means greater water-side head loss but lower flow rate. Looking at a very high temperature rise in coils has proven especially effective in desert areas with very high dry-bulb air temperatures, relatively low wet bulb, and a large percentage of outside airflow. As an example, some buildings at the University of California (UC) at Riverside with 100 percent outside air have as high as 80°F return chilled water temperature during the hottest outside air temperatures. UC Merced is designed with all the coils to provide return chilled water within 10°F of the inlet dry-bulb temperature. The thermal storage system at UC Merced operates on a blend of chilled water return from 100 percent outside-air (OSA) coils and building coils with return air, where the possible chilled water return temperature is less than that in 100 percent outside-air buildings. Most buildings at UC Merced achieve more than a 30°F temperature rise during the warmest weather. California State University (CSU) in San Bernardino often achieves 30°F delta-T employing a "most open valve" pumping strategy.

Colder chilled water supply and warmer chilled water return each contributes to smaller required thermal storage volumes for the same thermal storage capacity, less pumping energy used, and smaller pump and piping main requirements for the same cooling load.

Coil performance depends on several factors including the entering air wet- and dry-bulb temperatures, and the airflow over the coil. These two factors are key variables in how much cooling a coil is to provide. The higher the air dry-bulb temperature entering the coil, the higher the return water temperature leaving the coil can be. The cooling load on the coil is greatest when the entering air total heat is the greatest. With respect to dry-bulb temperature alone, this means that greater leaving water temperatures are possible when the entering air temperature is highest. However, the amount of moisture in the air can complicate this potential. When there is heavy latent load (moisture removal) on the coil, it is possible to have a greater cooling load at lower entering dry-bulb temperatures. Because of this condition, the best candidates for high leaving chilled water from coils are 100 percent outside air coils in a hot desert (dry air) climate. In those conditions, a reasonable coil design can achieve leaving return chilled water within 10°F of the entering air dry-bulb temperature. But with higher latent loads (return air, wetter climates, or monsoon conditions), full cooling load may occur at lower entering air dry-bulb temperatures, which will reduce the leaving chilled water return temperature and the capacity of the thermal storage system.

Preventing Increased Supply Water Flow

A higher chilled water supply temperature than design results in loss of cooling capacity or in increased water flow, if a variable-flow delivery system is used. Increasing the chilled water supply flow when chilled water return temperature lower than design is produced, and mixing the supply and return in the thermal storage tank to further increase the supply temperature, will result in an ever increasing reduction of the thermal storage system capacity. Thermal storage capacity depends on maintaining the proper design temperature rise (delta-T) for the water in the storage tank. If mixing raises the supply water temperature, that results in lower return water temperature and greater flow, as discussed above, with reduced cooling ability. Blending supply water with return water will also reduce the return temperature in the tank compared to that which occurs in the coils. The combination of these factors will result in a substantial thermal storage capacity loss.

Effect of Blending on Chillers

Reduced chilled water return temperature is interpreted by the chiller controls as an indication of low load and the chiller output will be reduced. For example, if a system uses a constant-flow design through the chillers with a return chilled water design temperature of 60°F and the actual return water temperature is 50°F, this will indicate low load to the chiller and the capacity controls will reduce the chiller output. If blending in the thermal storage tank artificially produces a lower return water temperature to the chillers, the chillers will respond by operating at lowered capacity.

Stratification System Design Requirements

Stratified thermal storage has certain requirements, most of which have been addressed in the preceding paragraphs. In addition to those items, several items that are necessary in successful stratified storage design are addressed below.

Proper Vertical Separation

Because stratified storage depends on differences in density to keep warm return water from mixing with cold supply water; the proper relationship of chilled water return and chilled water supply is essential to proper operation. Several unsuccessful chilled water storage systems have been constructed with simple pipes installed halfway down into large reservoirs on both the supply and return end. Obviously, such a design will produce mixing. One way or another, the lighter warmer chilled water return flow diffuser must be located at the top of the tank for the stratification to be maintained.

Diffuser Design

Proper diffuser design is one key to developing good stratification in the thermal storage tank. Water should be introduced in a thin layer, uniformly over the area of the tank. The velocity out of the diffusers must be very low to assure that the flow remains in the laminar flow region. Various suppliers of field-erected storage tanks provide stratified thermal storage tanks as "turnkey" (design-build) installations, for which the tank supplier takes responsibility for the design and installation of the diffusers, and guaranteeing the thermal performance of the storage tank; this approach has become routine since the mid-1980s. However, individual custom-designed diffusers are still used in some projects.

Reasonable Storage Time

Because heat is transferred across the interface between warmer chilled water return and the colder chilled water supply, the amount of time the tank will sit idle should be held to a minimum (usually less than a day). This is because, even though water is a very good insulator (four times better than ice, which is a fairly good insulator), water still conducts heat. In addition, any heat gains to the thermal storage tank can set up unwanted mixing (convection) currents within an idle tank. Nevertheless, large stratified chilled water storage tanks in seasonally variable climates are often recharged only once every 1 or 2 weeks, during periods of extended cool weather with light cooling loads; a benefit is that those light loads can be met 100 percent via pumping chilled water from storage (whereas, in a nonstorage situation, chillers would otherwise have to be operated at inefficient low part-load conditions).

Storage Below 39°F

Designers should remember that the change in density of water decreases as temperatures approach 39.4°F. At that temperature, pure water is its most dense (heavy) and actually gets lighter below 39°F. This does not mean that storage systems cannot be designed with lower than 39°F supply water. It does mean, however, that return temperature must be much higher in contrast for a density difference. When temperatures are close, slightly warmer water in the interface may be denser than the colder supply water and will tend to sink down into the supply and cause some mixing until the maximum density point of 39°F is reached. However, special aqueous fluids (water with chemical additives) are available that allow fluid stratification below 39°F, by reducing the temperature of maximum density (and the freezing point) for the water solution, such as SoCool® LTF. Some such LTF-stratified storage applications have been in service since the mid-1990s; the LTF storage has a typical operating supply temperature at or near 32°F, with some installations being as high as 36°F and as low as 30°F, while even lower temperatures are also possible.

Typical Stratification Challenges

Stratified thermal storage systems must include the above design basics to be successful. Designs that have *not* been successful had one or more of the following design problems:

- Improper tank flow design
- Constant flow chilled water system (and three-way control valves at the loads)
- Water flow in coils when air handlers are off
- Low delta-T design or operation
- Poor tank flow diffuser design
- Improper interface of tank, chillers, and building use systems

- Variable chiller supply temperature
- Lack of pressure control on coil return piping from high buildings

Good Stratified Storage System Characteristics

Conversely, there are many hundreds of very successful systems, including many that have operated well for 20 years, or more. Factors that can be identified as part of those designs are as follows:

- High delta-T design
- Variable-flow chilled water system
- Constant-flow chillers, often configured in series or series-parallel
- Tall vertical tank or similar physics
- Demand-based or time-based chiller control
- Two-way valves controlled at air handlers
- Chilled water flow off when air handlers are off

Serendipity of a Good Chilled Water Thermal Storage System

Serendipity is an unexpected benefit ancillary to the primary reason for an action. When a good thermal storage system is designed, the owner receives side benefits because those things necessary for good storage are good for any chilled water system. Some of the items to consider are outlined in the paragraphs below.

Reduced Chiller Energy Consumption

A good storage system can provide better energy performance than a conventional instantaneous system because of the high delta-T and its beneficial effect on series chillers, and because lower wet-bulb temperatures at subsequent lower condensing temperatures are available at night. Increased efficiency also results from all the chillers (and their auxiliaries) operating closer to design points more of the time.

Lower Pump Horsepower

Thermal storage designs have lower chilled water pumping energy needs than conventional designs because the high temperature difference required reduces chilled water flow requirements. Variable-flow systems circulate water proportional to the cooling; half the flow, in theory, only requires one eighth of the full-flow horsepower.

Smaller Chilled Water Pumps and Pipes

Smaller chilled water pumps and pipes can be used because of the lower flow of the high delta-T systems. This saves money on pumps, piping, and insulation costs. Where piping exists, more cooling capacity can be transported in the same pipe size.

Uninsulated Chilled Water Pipe Potential

A direct-buried uninsulated chilled water return pipe often may be used with high chilled water return temperatures. Good chilled water systems often develop return water temperatures within a few degrees of ground temperatures. Variable-flow systems do not return unused chilled water, which would cause condensation potential. These factors can combine to eliminate the need for chilled water return pipe insulation on high delta-T systems.

Better System Balancing and Control

Better system balance and control results from a variable-flow system, which automatically provides the system resources to the point of need. Because there is no waste of resources in blending, bypassing, and reheating unneeded cooling, there exists a very good ability to deliver those resources when and where they are needed. Variable-flow systems are ordinarily self-balancing and almost like a tap one turns to obtain water where and when needed.

High Reliability and Emergency Cooling

Higher reliability can be achieved with storage systems because they do not depend on instantaneous cooling production. Usually if a chiller is unavailable because of a power outage or failure of a component of the system, it is possible to serve critical areas for a period of time from thermal storage while restoring production. An example might be a computer room with critical cooling requirements. Dedicated pumps serving those critical areas might operate on an emergency generator and possibly provide capacity for days if the ratio of storage to load is right, and if the tank has chilled water storage when the failure occurs.

Lower Maintenance Cost

Maintenance costs are lower because thermal storage systems typically use smaller chillers, cooling towers, and pumps than do conventional systems. Although increased hours of operation may offset these savings to some extent, most maintenance is not impacted by hours of operation but rather by the number of stops and starts. In addition, typically, a conventional system runs most of the year in partial load conditions, which is just as demanding on maintenance as full-load operation.

Operational Flexibility

By decoupling the production of chilled water from the use of chilled water, storage provides operational flexibility. Inefficient, low part-load chiller operation need never be done.

Balancing Thermal and Electric Loads

Cool thermal storage flattens and balances daily load profiles of both cooling and electric loads. This can dramatically improve the economics and increase the feasibility of Combined Heat & Power installations, which in turn can provide substantial benefits in terms of economy, energy efficiency, and emissions.

Dual-Use as Fire Protection

Chilled water storage often serves a secondary function as an emergency fire protection water storage reservoir, which can reduce owners' risks and sometimes reduce insurance premiums.

Water Treatment

At least one particular LTF solution provides stable, long-term inhibition against both corrosion and microbiological activity, eliminating the need for typical on-going water treatment programs.

Reference

1. Brown, C. Z., *Sun, Wind and Light*, John Wiley & Sons, 1985, 130.

CHAPTER 5

Latent Thermal Storage Systems

Introduction

Most materials change temperature with the addition or removal of energy, except at one temperature point for a given pressure (which is unique for each material or substance) all the energy added or removed goes to a change of state between liquid and solid. The amount of energy required for a change of state between liquid and solid (latent heat of fusion) is very large in comparison to the energy required to change a material's temperature, 1°F or °C (sensible heat).

This large amount of thermal energy absorbed and released as a substance melts and freezes makes phase change materials (PCMs) excellent candidates for latent thermal storage applications. The most common phase change material employed in the thermal storage industry today is water. Several factors combine to make water the odds-on favorite for latent thermal storage. First, water is one of the most common and inexpensive materials available. Additionally, water is easy to transport, reasonably inert, and is easy to contain. Water's change of phase temperature is well suited to thermal storage for air conditioning applications. Most important is the amount of energy available in the change of state.

Therefore, it is not surprising that water in the form of ice as a thermal storage medium has an uninterrupted history dating back thousands of years. As discussed in Chap. 1, most people associated with the thermal storage industry are familiar with the centuries-old technique of harvesting ice from frozen bodies of water and mountainous areas for use in subsequent seasons. The practice dates back to a number of early civilizations, including the ancient Persians, who, by 400 BC successfully stored ice in "Yakhchal," thick walled, cone shaped structures formed from an insulating mortar.[1] Although far outnumbered by mechanical diurnal systems, there are still contemporary examples of seasonal ice storage.

Through a remarkably fortunate coincidence of nature, water, one of the commonest materials on earth, is uncommonly fitting as a PCM for cool storage. In addition to the characteristics that make water an ideal sensible heat storage medium, water also possesses a remarkably fortunate array of physical and chemical properties that provide an exceptional latent heat storage material. First, the melting point of 32°F (0°C), far above what would normally be expected for similar molecules, is well suited for comfort

---

This chapter was written by Brian Silvetti.

cooling, as well as for many process applications. The polar structure of the water molecule also results in intermolecular hydrogen bonding that contributes to the unusually high latent heat of fusion of 144 Btu per pound (Btu/lb) or in metric units 80 calories per gram(c/g). The combination of high latent heat, and relatively high density, endows water-ice with a high volumetric thermal storage capacity in excess of 8000 Btu per cubic foot (Btu/ft³). Another atypical property of water is the well-known characteristic of reduced density in the solid phase compared to the liquid phase—ice floats in liquid water (whereas the majority of materials are more dense as a solid than as a liquid). Although sometimes a challenge, this density property is sometimes an advantage.

The ice storage industry has developed a number of different technologies over the past several decades in an effort to find the best ways to take advantage of the significant benefits of ice as a storage material, while minimizing the disadvantages. Different approaches continue to emerge, while others have fallen out of favor. These ice storage systems generally fall into one of five or six basic categories, with a variety of additional individual design features. A consensus in descriptive nomenclature has gradually evolved for the basic system types that are described in this chapter.

Ice Thermal Storage Technologies

Ice-on-Coil

As discussed briefly in Chap. 3, one of the most widely applied basic systems is referred to as "ice-on-coil," which includes two primary subcategories that are defined by the discharge, or melting, process, either "internal melt" or "external melt."

First, before examining the different ice storage categories in more detail, the following pages describe the available ice-on-coil storage products in general, concentrating on ice production. The common element in the ice-on-coil category is the use of a heat exchanger(s) immersed in a tank of water. Cold fluid, circulating through the heat exchanger, freezes the water on the external surface of the heat exchanger. Since the fluid must obviously be lower in temperature than 32°F, an antifreeze solution (secondary coolant), or direct refrigerant, is used. Refrigerants are used only with metallic heat exchangers, while coolants are applied to all heat exchanger materials, including plastics.

Much of the popularity of secondary coolant-based storage is attributable to the coolant's compatibility with conventional chilled water system design and equipment. While different ice storage systems require somewhat different coolant temperatures to produce ice, all common types of central plant, vapor compression chillers are utilized, including the following: centrifugal, reciprocating, rotary (screw), and scroll. Additionally, the ice storage heat exchangers simplify the integration into positive pressure, closed loop, chilled coolant systems.

When an antifreeze solution is used to produce the ice, the solution or fluid experiences a temperature change as it travels through the ice storage heat exchanger, entering at a lower temperature and gradually approaching 32°F as it nears the exit of the heat exchanger. Ice will form at a faster rate where the fluid enters the heat exchanger since the fluid is colder and there is relatively larger temperature difference. Manufacturers often attempt to arrange their heat exchangers so that flow in adjacent tubes will be in opposite directions, insuring that the maximum volume of water participates in the storage process. In addition, the same effect occurs during the melting process for the internal melt equipment described below.

Figure 5-1 reproduces the actual, ice-making record of the entering and leaving coolant temperatures for a modular ice-on-coil thermal storage tank with a plastic tube heat exchanger. The temperature changes follow the characteristic profile of a material experiencing a change in phase from liquid to solid. The initial steep temperature drop corresponds to the reduction in temperature of the liquid water before ice forms in the storage tank. When ice begins to form, the thermal energy extracted from the water results in phase transition to solid ice with little change in temperature, and after most of the water has been frozen the temperature once again begins to decrease more rapidly. As ice is formed, the temperature of the coolants slowly decreases due to the gradual accumulation of ice around the heat exchanger tubing (Fig. 5-2); the 32°F surface exists at the solid/liquid interface, which is gradually receding from the heat exchanger. However, the coolant leaving temperature is within a few degrees of the freezing temperature throughout the charging process.

Note the feature at the beginning of the ice-making temperature plateau, where the temperatures decrease a few degrees before suddenly rising to a more constant level. This phenomenon, known as supercooling and discussed in the section on salt hydrates, demonstrates the tendency of many materials, including water, to remain in a liquid state, even below their normal freezing point. While water can theoretically be supercooled to about −40°F under carefully controlled laboratory conditions, the storage tank environment will induce nucleation as shown within only a few degrees of the customary 32°F.

A refrigerant evaporating within metallic heat exchangers is the other method of ice production. Utilizing refrigerants directly within the storage heat exchanger possesses a perceived advantage, in that it eliminates the intermediate level of heat exchange

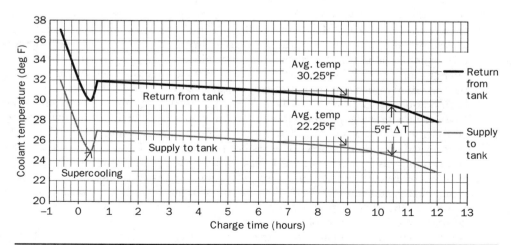

Figure 5-1 Supply and return coolant temperatures of an internal melt, ice-on-coil modular thermal storage tank.

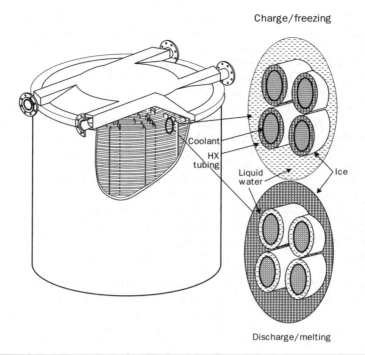

Charge/freezing

Coolant
HX
tubing

Liquid
water

Ice

Discharge/melting

FIGURE 5-2   The formation and melting of ice in an internal melt, ice-on-coil thermal storage tank.

between the secondary coolant and the ice. And in fact, this method has proven successful in smaller capacity equipment and in products for many processes. Because this approach allows the compression refrigeration evaporator temperature to approach the phase change temperature of the water-ice storage, this process can be more efficient than those that must cool an indirect fluid. There are, however, disadvantages in the large amount of refrigerant needed, special equipment required, and the difficulties or challenges of handling the refrigerants and lubricants. The proper distribution of an evaporating refrigerant throughout an extensive matrix of heat exchangers, while safely managing oil return to the compressor, is quite challenging and was usually addressed with liquid overfeed, where refrigerant is pumped through the heat exchanger at two to three times the evaporation rate. In addition to other problems discussed later, the simplicity of conventional chilled water design practices, using off-the-shelf chillers, pumps, valves, and other components has proven more popular in the larger capacity commercial environment. There is also additional emphasis today on limiting the volume of refrigerant contained by a system (in order to limit leakage into the atmosphere), and commercial chillers are frugal in this regard. All of these reasons have resulted in the larger systems found in commercial comfort cooling applications largely adopting the secondary coolant approach.

The ice-making heat exchangers are typically serpentine or spiral tubular arrangements. Products designed for larger commercial chilled water systems utilize plastic or steel tube heat exchangers with copper appearing in smaller capacity equipment intended for the unitary cooling market. The individual tubes of the heat exchangers extend throughout the internal volume of the tanks, with assemblies of interconnected heat exchanger modules often found in larger, site built tanks. The spacing and arrangement of the heat exchanger tubes are carefully engineered to freeze and melt (for internal melt equipment) the maximum quantity of water, at the proper rates of heat transfer, while displacing as little of the tank volume as possible. Current commercially available products are all designed to charge (freeze) and discharge (melt) within daily cycles.

An additional objective is to minimize the temperature difference between the circulating coolant and the phase change temperature of the storage medium. Charging (ice building) with relatively high temperatures increases refrigeration system efficiency and capacity, and melting with low coolant temperatures can reduce required system pumping energy and enhance heat transfer capability throughout the cooling system.

Internal Melt Ice-on-Coil

As noted, the major subdivisions of ice-on-coil storage equipment are defined by the method used to melt the ice during the cooling mode. Internal melt designs employ the same heat transfer fluid flowing through the heat exchanger that was used to previously produce the ice. The majority of comfort cooling capacity currently being installed utilizes an antifreeze liquid, secondary coolant as the heat transfer fluid. Refrigerants (e.g., R-410a) have also been employed as the discharge heat transfer fluid but currently are usually found only in smaller capacity equipment in the range of 40 ton-hours. These small unitary products include the refrigeration system with the storage tank as a packaged combination.

As the ice changes back to the liquid phase in an internal melt system, an annulus of liquid water forms between the heat exchanger surface and the remaining ice as shown in Fig. 5-2. As more ice is melted, the intervening layer of liquid water gradually increases in thickness. Initially, when ice is close to the heat exchange surface, the heat transfer fluid exiting the storage tank heat exchanger approaches 32°F, the phase change temperature of water. As the annulus of liquid water grows, the delivered coolant temperature gradually increases due to the insulating properties of the liquid water layer. In order to maintain the required cooling capacity delivered by the storage system, flow through the storage heat exchanger is gradually increased to compensate for the rise in the coolant leaving temperature. Quite often, managing the flow is accomplished by an automatic three-way valve arrangement that directs the necessary volume of coolant flow through the storage heat exchanger needed to maintain the proper temperature delivered to the cooling load, which can also be accomplished in some designs with a variable speed pump.

Another technique occasionally employed to augment heat transfer performance in the discharge mode is to induce convective motion of the water within the tank by introducing compressed air or, in the case of the smaller capacity equipment, recirculating melted water. However, air agitation is more commonly associated with external melt products.

While the discharge performance of internal melt systems can decrease as more ice is melted, when the system returns to the ice-making mode, the ice storage system always first freezes water directly in contact with the heat exchanger, maximizing the efficiency of the ice-making process. For a variety of reasons, internal melt, ice-on-coil systems are often provided as factory manufactured modular products. Relatively compact units, suitable for over-the-road shipping, possess substantial capacity due to the inherently high energy density of ice, making a modular approach practical for many applications. Several basic system designs are available including thermoplastic heat exchangers within metallic, rotationally molded plastic or fiberglass tanks. Steel heat exchangers for the modular products are usually found in galvanized steel tanks but they could also be located in large site-built concrete tanks. The copper heat exchangers found in the smaller unitary products typically also employ plastic tanks.

Because the modular storage tanks are completely factory manufactured products, they are often provided with performance data, computerized selection assistance and factory engineering support. The modular tanks can be installed in almost any location— mechanical rooms, basements, outdoor pads, rooftops and some may even be buried. Installations can be found in small churches, office buildings, schools or large campus and district cooling applications. Besides the intrinsic redundancy afforded by modular systems, modular systems can also be applied to a very broad range of system capacities. Very similar modular products can be seen in both a small 320 ton-hour (1125 kilowatthour (kWh) thermal) school and a large 13,000 ton-hour (45,700 kWh thermal) district system in Fig. 5-3.

External Melt Ice-on-Coil

External melt systems do not use the same fluid to melt the ice as was used to produce the ice. Rather, some liquid water is always preserved between the layers of ice forming around the heat exchanger pipes or tubes and the storage tank. This water is circulated to the cooling coils or intermediate heat exchanger, where it gains heat and returns to

FIGURE 5-3 Similar modular storage tanks in a 340 ton-hour school installation and a 13,000 ton-hour district cooling system.

the ice storage tank to once again be cooled by direct contact with the remaining ice, melting some of the ice in the process.

In most thermal storage systems, it is desirable to use the ice-producing refrigeration equipment to help meet cooling load during the discharge mode. While it is possible to just continue operating in the ice-making mode, for external melt products, this would require that the refrigeration equipment continuously run at ice-making temperatures since ice is always in contact with the storage heat exchanger. Commonly used alternatives, intended to increase the direct cooling efficiency, include redirecting ice-making refrigerants or coolants to a separate heat exchanger where they operate at the higher, and more efficient, daytime system temperatures.

The direct contact heat exchange in external melt is capable of achieving very high rates of discharge and reliably low temperatures that approach 32°F throughout the discharge. External melt products are available as individual modular storage tanks usually for process, dairy, or agricultural applications; however, the larger commercial cooling equipment will more typically be designed as field erected, rectangular concrete tanks with many individual heat exchangers plumbed together as shown in Fig. 5-4. An external melt installation in New Orleans stores 60,000 ton-hours of cooling capacity in a tank that measures 77 ft long, 61 ft wide, and 38 ft deep.

The heat exchangers are arranged so that the unfrozen passages are oriented vertically, extending from the tank bottom to the top and through the entire tank length. The tanks are usually quite large and operate at atmospheric pressure. Like other open systems, it is sometimes necessary to interface with pressurized building loops or distribution systems at different elevations.

Since it is essential that passages through the ice are always unobstructed, careful control of the ice thickness is crucial. Warm water entering the tank will melt the most ice so ice thickness along the flow path of the tank will not be constant. Careful monitoring of the ice condition, through sequences of both charge and discharge, is

Figure 5-4 Galvanized steel, external melt, thermal storage heat exchanger lifted into site built tank.

essential and complete melting of the ice inventory may periodically be employed to restore the initial uniform state.

The velocity of the water through the thermal storage tank in the melting mode is very low; and it is difficult to insure well-distributed flow throughout the tank volume; and, therefore, it is customary to supply compressed air through a matrix of plastic piping installed below the lowest level of heat exchangers. The vigorous agitation provided by the rising air bubbles helps promote both even melting and freezing, although the agitation may not be active through the entire ice-making period.

Although external melt storage is found in all types of cooling applications, external melt storage has been particularly successful in district cooling systems, perhaps the largest being the 300,000+ ton-hour system installed in five plants, that cools 100 buildings in downtown Chicago.

Encapsulated Ice Storage

Encapsulated storage systems possess some similarities to internal melt, ice-on-coil technologies, which also rely on secondary coolants to freeze and melt water, and the phase change water is not circulated. The water is sealed in containers that are then loaded into tanks. The coolant circulates through the tank and around the individual containers of water.

Encapsulated thermal storage tanks can be virtually any shape or size. Although encapsulated thermal storage tanks typically use horizontal or vertical steel cylinders, concrete construction is also employed. Steel tanks operate at atmospheric or system pressures, but concrete tanks are usually limited to atmospheric pressures. Single or multiple tanks can be located outdoors, inside or buried.

The individual encapsulated ice containers are usually constructed of thermoplastics. Plastic is easily formed into any desired shape, is inexpensive, corrosion resistant, and exhibits excellent toughness at low temperatures. The dimensions of the containers must be carefully selected so that all or most of the internal water undergoes a change of phase while providing acceptable heat transfer rates and coolant temperatures within the appropriate time periods. A number of different shapes have been available over the years. Flat trays were a logical choice since they could be designed to stack closely together, making the best use of the available tank volume. However, most systems today employ spherical shapes, typically 3 to 4 inches in diameter. Some containers incorporate surface sections that can alternate between concave and convex shapes to accommodate the volume change between the solid and liquid phases. Others simply rely on the inherent strength of the spherical shape and some void space within the container. There were some early attempts to provide containers that could be filled with water and sealed on-site. However, the current preference is to take advantage of the reliability and quality control provided by manufacturer in a factory setting. Any perceived savings in shipping cost is largely illusory since the volume of the product is the same.

Containers that adjust their shape to accommodate the volume difference between solid and liquid will displace coolant volume as the water freezes. An additional vessel is needed to accept the displaced liquid. However, the amount of displaced coolant can be used to indicate the inventory of stored cooling.

System design with encapsulated storage is similar to internal melt, ice-on-coil. The storage is easily integrated into conventional chilled water cooling system designs, and ice can be produced using the same types of conventional chillers. Moreover, for similar reasons, the ice storage can be located upstream, downstream, or in parallel with the ice-making chillers. A secondary coolant is used to both freeze and melt the ice in the thermal storage tank. Additionally, the potential discharge rate and achievable coolant temperatures vary with the ice storage inventory. However, due to the different physical geometry of the variation in heat transfer path, the trajectory of the coolant temperatures and discharge rates will be quite different from those for internal melt ice-on-coil devices. On discharge, the entire internal surface of the water containers is initially in direct contact with ice. As ice is melted, a layer of liquid water forms between the mass of ice and the container surface, but because ice is less dense than liquid water, some will float within the container, maintaining some direct contact with the surface.

Because encapsulated systems store the water permanently within a completely sealed container, encapsulated ice system technologies can be applied to alternative PCMs; and some manufacturers catalog a variety of available materials with a broad range of fusion temperatures. In practice, however, water is the almost universally applied material.

Encapsulation is more likely to promote supercooling; and thus manufacturers will include proprietary nucleating agents within the capsules. Other applications and methods of encapsulation are beginning to appear and are discussed under "Other PCMs."

Ice Harvesters

Ice harvesters are quite different from any of the technologies described so far. Ice harvesters were developed to address two phenomena in the production of ice. First, when water is frozen on a heat exchange surface, it forms a strong bond between itself and the heat exchanger (evaporator). Second, the ice layer forms or acts as an insulator, which reduces the rate of heat exchange between the refrigerant and the water being frozen. With ice harvesters, water constantly recirculated from a tank below falls as a film over vertically oriented plates or tubes that are often constructed of stainless steel. These plates or tubes are essentially refrigerant evaporators, and the water freezes in thin layers of ¼ to ½ in. thickness directly on the evaporator surface. There is no intermediate coolant as used in most internal melt and encapsulated ice systems. Periodically, typically at intervals under 30 minutes, refrigerant valves redirect hot gas from the compressor discharge into the evaporator plates or tubes, releasing the ice from the evaporator surface, allowing the ice to fall by gravity into a storage tank below. Multiple plates or tubes can be employed so that the process is continuous, with some surfaces collecting ice while others are harvesting or shedding the ice. The ice usually fractures into smaller pieces as it falls, and the process can be mechanically assisted, depending on the thickness and shape of ice produced.

R-22 and ammonia refrigerants have been staples of the industry for many years, but others, such as R-404a or even propane, are available depending on the equipment design.

It is interesting to note that, as ice accumulates in the tank of water below the harvester, the liquid water level does not change until enough ice has accumulated to a point where ice reaches the bottom of the tank and actually supports ice above. When ice floats in liquid water, it displaces its weight, which is exactly equal to the weight of the volume of liquid from which the ice was produced. Since ice has about 10 percent more volume than water of the same weight, about 10 percent of ice floats above the water level.

The ice harvester refrigeration systems are relatively expensive in comparison to the cost of the thermal storage tank. Therefore, ice storage is often accumulated on a weekly cycle rather than the more common diurnal or daily cycle. The refrigeration system can run continuously, including weekends, reducing the required refrigeration and ice harvester capacity while increasing the (less expensive) storage tank volumes.

During thermal storage discharge, chilled water is circulated from the bottom of the storage tank to the cooling load, returning to flow over the harvester evaporators. During the discharge, harvester evaporators can operate as a chiller upstream of the storage, contributing cooling capacity in a partial storage mode. There are some limitations however, as it is not practical to divert the flow around the storage tank after it flows over the evaporator. Ice will be melted anytime the temperature of the falling water is greater than 32°F, even if it is preferred to carry the cooling load entirely with the harvester operating as a chiller at a higher water leaving temperature.

On the other hand, as noted, the direct contact of the circulating water with a very large ice surface area provides chilled water temperatures that approach 32°F throughout the discharge with dependably high discharge rates (as long as no "short circuit" paths of open water have developed in the ice matrix).

Ice harvesters are obviously open systems operating at atmospheric pressure. Like ice-on-coil external melt designs, measures must be taken to interface with a pressurized distribution system that may also be at a different elevation, with such measures being, for example, pressure sustaining valves or a heat exchanger between the storage system and the building cooling loop. Ice harvesters are quite common in process, agricultural, food preservation, and other industrial applications. Some very large systems have been applied to early combustion turbine inlet-air cooling applications. One of the earliest is the 173,000 ton-hour ice harvester and storage tank that cools 930,000 lb of air per hour that flows into eight 2-megawatt combustion turbines in Fayetteville, NC.[2]

Ice Slurry

Ice slurry equipment bears some similarities to ice harvesters. Water freezes as it flows over a refrigerated surface, usually the interior face of a cylinder. However, a hot gas defrost cycle is not needed to separate accumulating layers of ice. Rather, a low concentration of a freeze depressant is added to the water. As covered in more detail in the discussion on other PCMs, the freeze depressant comes out of solution as pure water freezes. This process limits the growth of the ice crystals and also helps to prevent adhesion of the ice to the refrigerated surfaces. Mechanically activated scrapers, rods, or screws augment the separation of the crystals and sweep them away from the cylinder walls to be carried to a separate thermal storage vessel.

A variety of freeze depressants can be used including the following: ethylene or propylene glycol; alcohol; sugars; and salts, such as sodium chloride.[3] Initial concentrations vary from just a few percent to 10 percent, depending on the application and freeze depressant.

There are several options for slurry thermal storage and use of the ice slurry itself. An often cited potential advantage of ice slurries is the opportunity to distribute the slurry directly to the cooling application. Because of the large amount of heat capacity provided by the change of phase, pump energy, pipe sizing, and distribution energy could be dramatically reduced. Although there are examples of this practice, coagulation of the ice somewhere within the piping system is a possibility. The fraction of ice in the slurry must be relatively low, and the storage vessels are typically constantly mixed by mechanical stirring. The distribution system must be carefully designed, and the lower ice fractions increase storage tank size. Higher ice fraction slurries can be treated in a manner similar to ice harvesters. Warm water returning from the cooling load is sprayed over the top of the slurry and chilled water is extracted from below, which is the most common approach employed in space-conditioning applications. A third method mechanically removes ice from the top of the storage tank where it has separated from the brine by gravity. The ice is then blended with additional water in a separate tank and often used for direct cooling of food products such as for the fishing industry.

Another ice slurry technology operates at the triple point of water, where all three phases (solid, liquid, and vapor) are in equilibrium. This occurs at a temperature of 32°F and a low absolute pressure (strong vacuum). A vapor compressor (vacuum fan) withdraws water vapor from the ice generation vacuum vessel, where some of the liquid water is boiled (at a heat of vaporization of about 1000 Btu/lb), while some of the liquid water is frozen (at a heat of fusion of 144 Btu/lb); thus, approximately 7 lb of ice is formed for each 1 lb of water vapor boiled. The ice crystals generated form a pumpable ice-water slurry. The ice slurry can be stored in a remote tank at concentrations up to about 50 percent by weight; and the ice slurry has been pumped in concentrations up to about 20 percent by weight. The use of water vapor as the refrigerant provides the benefit of no heat transfer surface at the point of ice generation, as well as an elevated refrigeration compressor suction temperature for reduced energy consumption compared to traditional ice-making technologies. A trade-off is the relatively high capital cost of the vacuum vessels and especially the high specific volume vacuum compressors. This technology of triple-point ice slurry generation, originally developed in Israel in the 1960s as a means of seawater desalination, has been used in the 1990s and 2000s in varied applications, including ice production for deep mine cooling in South Africa, heat pump application for district heating in Scandinavia, all-weather snow making in European ski resorts, as well as thermal storage for space conditioning.

Seasonal Ice Storage

While comparatively rare, seasonal storage systems continue to be applied. One notable example is the regional hospital in Sundsvall, Sweden, which presently has been in operation about 10 years.[4] Up to 40,000 m³ (1,413,000 ft³) of snow is collected and produced over the winter, with up to 70 percent of the snow generated with "snow-guns," the remainder naturally produced and collected by the municipality. The snow is covered with a 0.2 meter (8 in.) layer of wood chips which helps reduce annual losses from heat gain to about 20 to 30 percent of the total capacity.

The total delivered quantity of cooling varies from year to year but has reached a maximum of 1345 MWh (382,400 ton-hours) with 87 percent of the facility cooling loads provided by the snow. The snow does not typically last the entire cooling season and an 800-kW (225-ton) chiller meets the cooling load when the snow has been exhausted or when the snow pumping system is inadequate for the cooling load.

Ice Storage: Secondary Coolants

A key component of both ice-on-coil and encapsulated storage systems is the fluid that circulates to build and melt ice. Virtually all of the fluids that are in current use are water based, with a range of freeze point depressant additives, the most common being a solution of ethylene glycol and water, similar to familiar automotive antifreeze. Other fluids have been, and continue to be, used where a freeze-point depressant is a requirement. Inorganic salts, such as calcium chloride, have been a staple of low temperature systems such as skating rinks for many years. However, inorganic salts are not often found in the commercial cooling environment and in fact, many facilities have converted to glycols. More recently, organic salts such as formates, acetates, and related combinations have begun to achieve some level of acceptance. Alcohol, particularly methanol, was a popular antifreeze during the early part of the 20th century. While still occasionally applied, it has been largely displaced by ethylene or propylene glycol.

While automotive solutions are typically a 50/50 mixture of water and glycol, fluids applied in thermal storage are usually in the range of 25 to 30 percent ethylene glycol. The use of lower concentrations of glycols is not recommended, due to the increased potential for microbiological activity. Although automotive antifreeze is glycol based, it should never be used in a thermal storage system application. A number of manufacturers produce heat transfer grade ethylene glycols that include a carefully formulated corrosion inhibitor package specifically designed for the long life, concentrations and materials found in central cooling/heating plants, and without undesirable automotive grade additives such as silicates.

Glycols have good chemical stability and possess acceptable heat transfer and fluid flow properties. Glycol vapor pressures are low; and diluted glycols are essentially nonflammable in the concentrations typically employed for thermal storage applications. Glycols are compatible with most materials employed in commercial mechanical systems, except galvanized surfaces; and glycols have reasonable cost.

Note that pure ethylene glycol does not offer the corrosion protection that formulated, commercial heat transfer grade glycol products provide. In fact, pure glycol is more corrosive in many ways than untreated water. Many of the heat transfer fluid manufacturers offer free annual testing of the fluid and, if necessary, these fluids can often be reconditioned with corrosion inhibitor, rather than replaced. Commercial grade fluids have lifetimes that can reach 20 years and longer when applied at the relatively cold temperatures found in thermal storage systems for cooling applications.

Ethylene glycol is considered moderately toxic and should not be used where accidental ingestion or contamination of potable water is possible. Propylene glycol, which is more costly, is often substituted in these cases; but its higher viscosity results in higher pumping energy and slightly reduced heat transfer performance. In addition, propylene glycol freeze protection equal to that of ethylene glycol requires a slightly higher concentration.

If released to the environment, both ethylene and propylene glycol completely biodegrade, consuming oxygen in the process. Ethylene and propylene glycol are chemically compatible with each other, but they should not be mixed, as the properties used to determine concentration, such as density or refractive index, no longer correspond to a unique percentage of glycol. Otherwise, the application and use of propylene and ethylene glycols are similar.

Other Phase Change Materials and Applications

Of course, materials other than water, as well as alternative phase transition mechanisms, have been used for latent (phase change) thermal storage applications, including some intended to replace water for cooling. Like water, PCMs absorb heat during the melting process and release heat during the freezing or solidification process. After decades of research, some of these materials are beginning to enter the marketplace,[5] often as a means to increase the thermal capacity of conventional building materials. Few of these PCMs are suitable as a heat transfer fluid in the liquid phase, dictating the use of a heat exchanger between the storage material and a distribution medium. Additionally, some of these PCMs, particularly the salt hydrates and eutectics, can be quite corrosive. Macro or micro encapsulation is the typical method of achieving the required separation.

Perhaps the family of materials that has received the most attention for thermal storage above 32°F is the salt hydrates. Salt hydrates typically have high density and high latent heat capacity at well-defined solid/liquid transition temperatures. Hydrates incorporate water into their crystalline structure; but the water is not chemically bound to the other molecular components, that is, H_2O remains H_2O throughout the liquid/solid phase transition. The water is integrated into the crystalline structure in exact molecular ratios.

Sodium sulfate is an interesting example, because its various embodiments cover a very broad range of potential thermal storage applications. For instance, one of the more extensively studied hydrates, sodium sulfate decahydrate (Na_2SO_4), Glauber's salt, combines 10 H_2O molecules with every 1 Na_2SO_4 molecule. With a phase change temperature of 90°F, Glauber's salt has been proposed as an ideal storage material for passive and active solar heat storage for over 50 years. Glauber's salt exhibits both the positive and negative attributes that characterize many salt hydrate storage materials.

Like many of the hydrate salts, sodium sulfate is relatively inexpensive, chemically stable, and benign. The decahydrate density of 91 lb/ft³ (1.46 g/cc) and latent heat of 104 Btu/lb [241 kilojoules per kilogram (kJ/kg)] provide a volumetric heat capacity theoretically on a par with water. In practice, the heat capacity is reduced due to the addition of critically important additives. The phase transition temperature is also often depressed slightly. Like most of the hydrates, Glauber's salt does not exhibit any significant vapor pressure, simplifying handling, installation, and application. Water loss or gain must be prevented, which also leads to encapsulation as the preferred installation method.

Unfortunately, Glauber's salt also presents many of the classic challenges often found in hydrate storage materials. Incongruent melting is perhaps the most difficult obstacle to overcome. When Glauber's salt melts at 90°F, about 15 percent of the sodium sulfate comes out of solution as anhydrous salt. Due to its higher density, this salt will settle to the bottom of the containing vessel, making it unavailable for recombination, unless measures are taken to hold the salt in suspension. Several different methods to hold the salt in suspension have developed over the years. These include storing the material in a continuously rotating drum or the addition of a thixotropic agent or gel. Natural clays, or colloidal and polyacrylamide gels have all been employed to keep the salts in suspension and in a position where it can recombine with the available water as heat is withdrawn and the material solidifies. Besides the loss of storage capacity, separation of salt from its water of hydration can create the opportunity for formation of unfavorable hydrate forms, such as the heptahydrate for sodium sulfate.

Another common characteristic of the hydrates, referred to as supercooling, is the tendency to persist as a liquid, even though the temperature is below the normal fusion temperature. If the material does not change phase, it does not yield its latent heat of fusion. The addition of a material with a similar crystalline structure, and higher fusion temperature, can provide sites for nucleation of the crystalline phase at the normal melting point. Another technique, referred to as a "cold finger," includes a thermally conductive path to a lower temperature location that will induce nucleation. However, this approach has obvious limitations; for instance, it is not effective for encapsulated storage systems.

Interestingly, the resistance to nucleation of the solid phase has been used to provide an on-demand heat source. Sodium acetate trihydrate, with a strong resistance to nucleation,[6] can exist as a liquid at room temperature, well below its normal fusion point of 136°F. A sudden, sharp disturbance of the liquid or mechanical shock can initiate the nucleation process, and the material rapidly returns to its phase transition temperature as it solidifies, releasing heat. Unlike irreversible exothermic chemical reactions, sodium acetate can be repeatedly regenerated by heating until completely liquefied. If any solid crystals remain, supercooling will not occur.

Various hydrates continue to be explored as thermal storage candidates.[7] Tetra-n-butylammonium bromide (TBAB), a quaternary ammonium salt, can be frozen to a slurry for cooling applications that do not require a multitude of tank-heat exchangers, since it can be pumped to a conventional refrigerated heat exchanger for heat addition and extraction. While the volumetric heat capacity is limited, it does have a higher phase change temperature than water, enabling chillers such as lithium bromide absorbers to produce stored cooling.

Similar to the salt hydrates are clathrates, or gas hydrates, in which gas molecules become caged within a water lattice. Clathrates received some measure of notoriety during the 2010 Gulf of Mexico oil spill when ice like methane hydrates formed within the "cap" placed over the rupture, blocking the flow of the petroleum being recovered by surface vessels. In addition to naturally occurring gas clathrates such as methane, propane, and ethane, refrigerant clathrates have also been proposed; but there has been little commercial development.

Hydrocarbons, such as paraffins or fatty acids are also common candidates for storage materials.[8] Paraffins are in the same chemical family as other familiar alkanes such as methane and mineral oil. Unlike the salt hydrates, these compounds readily nucleate, and they typically do not suffer from incongruency. As with hydrates, materials are available with a broad selection of phase change temperatures. Refined petroleum products are somewhat more costly than the salt- and water-based alternatives.

The hydrocarbons, however, usually have lower density and latent heats than the salt-based alternatives. For example, tetradecane (sometimes blended with other alkanes), with a phase transition temperature of 42.4°F, is a frequently proposed substitute for plain water/ice in cooling applications.[9] With a specific gravity of 0.763 and latent heat of fusion of 97.6 Btu/lb (227 kJ/kg), the volumetric heat capacity of these hydrocarbons is only about half of the value for water. Specific heat of the liquid at 0.52 Btu/lb-°F (2.18 kJ/kg-°K) is also about half that of water. For latent heat storage systems, sensible heat still contributes a significant fraction of the total capacity. The paraffins also suffer from poor thermal conductivity, many about one-fourth the value for water. Extended heat transfer surfaces and the blending of highly conductive additives are potential methods for overcoming the material's heat exchange limitations.

As with the salt hydrates, development continues in the development of organic PCMs, such as derivatives of natural vegetable oils and fats.

In addition to the familiar liquid/solid change of phase, other transformations of the molecular structure of a material are accompanied by the absorption or rejection of thermal energy. Returning briefly to the sodium sulfate example, a modest quantity of latent heat is associated with a change in the crystalline structure of the anhydrous salt at approximately 450°F, and organic materials can exhibit the same behavior. Hexatriacontane, for instance, goes through two solid-solid phase changes and a liquid-solid phase change within the range of 162 to 169°F. While an interesting phenomenon, practical applications are relatively uncommon.

For many years, these alternative PCMs enjoyed limited success in a number of specialized applications. Medical heat packs and wraps, constant temperature shipping containers, sports apparel and electronic component temperature stabilization are just a few of the products that occasionally appeared incorporating PCMs. However, after many years of aborted attempts, PCMs are beginning to address the higher thermal capacity, space-conditioning functions.

Many of these PCMs are selected with phase change temperatures near typical, conditioned space levels. As temperatures rise, the PCM absorbs heat, preventing overheating of the space. The heat can then be released when temperatures drop, maintaining more stable temperature conditions, with the PCMs essentially acting as thermal capacitors. In one embodiment, microencapsulated paraffins are infused into construction drywall,[11] conferring the thermal capacity of high mass construction in a lightweight material. In another application, a salt hydrate is contained within chambers formed by the layers of glass in a manufactured commercial window.[12] Cooling applications are also addressed by packages of PCMs designed to be placed in ceiling plenums or modified ceiling tiles.

The current interest in renewable energy will certainly intensify the development efforts of PCMs across a broad range of applications and temperatures.

Phase Change Below 32°F

Hydrocarbons and salt-based materials are available with phase change temperatures below the freezing point of water. The role of freeze point depressants such as ethylene glycol, or salts like sodium chloride, is the source of a common misconception regarding PCMs and the behavior of eutectics in general. Referencing a freeze point chart for solutions of sodium chloride and water,[12] a 10-percent concentration corresponds to a freeze point of about −6.5°C.[10] However, this is the temperature at which the first solid crystals of water form. In closed (fixed volume) systems, the salt that was in solution with the now frozen water is liberated, increasing the salt concentration of the remaining liquid, further depressing the freeze point. Rather than changing phase at constant temperature, a gradual drop in phase change temperature occurs until a "eutectic" concentration is reached, at which point all of the liquid solution will solidify at constant temperature. In the case of sodium chloride, the eutectic concentration is 23.3 percent, with a uniform melting point of −21.2°C (−6.2°F). A major benefit of phase change thermal storage is the large amount of energy that is available at a constant temperature, and the sliding fusion temperature of noneutectic concentrations provides little value. Much of the effort in thermal storage technology is focused on finding appropriate eutectic compositions and ensuring their long-term reliability. As discussed above, sodium sulfate and water forms a decahydrate composition with a melting point of 90°F and, in its anhydrous form, also exhibits a solid-solid transition at 450°F. However, at lower concentrations with water, sodium sulfate will also form a eutectic with a phase change temperature of 28°F.

Salt and water mixtures with phase change temperatures below 32°F are considerably easier to manage, principally because the salt concentrations are typically much lower than they are for the higher temperature hydrates and they usually work within their solubility limits. Separation of the salts is still possible, but corrective measures are simpler to implement.

Summary

While phase change thermal storage has a broad variety of potential applications, ice-based cooling storage has achieved, by far, the most success with respect to latent thermal storage systems. This success is primarily due to the ideal properties of water as a phase change material and the role of cooling loads in creating the strain on our electric grid. Very different technologies are available, including internal and external melt variations of ice-on-coil, encapsulated and different types of ice harvesters, as well as ice slurry. Ice storage has been applied to systems as small as a single residence and as large as district systems serving over 200 individual buildings.

References

1. Marks, G., *Encyclopedia of Jewish Food*, John Wiley & Sons, Hoboken, NJ, 2010.
2. Beaty, L., "Lincoln to Fayetteville: Bridging the Gap," *Proceedings of the Combustion Turbine Inlet Air Cooling Conference*, Fayette, NC, Aug. 18–19, 1993.
3. Wang, M. J., and V. Goldstein, in "Ice Slurry Based Thermal Storage Technology," *Proceedings of the IEA-Annex 17: 7th Expert Meeting and Workshop*, Beijing, China, Oct. 11–12, 2004.
4. Skogsberg, K., *Seasonal Snow Storage for Cooling Applications*, Lulea University of Technology, 2005.
5. Zalba, B., J. Marin, L. Cabeza, H. Mehling, Review on thermal energy storage with phase change: materials, heat transfer analysis, *Applied Thermal Engineering*, 2003, 23, 251–283.
6. Rogerson, M., and S. Cardoso, "Solidification in Heat Packs: II. Role of Cavitation." *AlChE Journal*, 2003, 49(2).
7. Ogoshi, H., E. Matsuyama, H. Miyamoto, T. Mizukami, N. Furumoto, M. Sugiyama, In "Clathrate Hydrate Slurry, CHS Thermal Energy Storage System and Its Applications," *Proceedings of the 2010 International Symposium on Next-Generation Air Conditioning and Refrigeration Technology*, Tokyo, Japan, Feb. 17–19, 2010.
8. Mehling, H., and F. Cabeza, *Heat and Cold Storage in PCM: An Up to Date Introduction into Basics and Applications*, Springer, Oct. 2008, 238–255.
9. Bo, H., M. Gustafsson, and F. Setterwall, in "Paraffin Waxes and Their Binary Mixture as Phase Change Materials (PCMs) for Cool Storage in District Cooling System," *IEA Annex 10, Phase Change Materials and Chemical Reactions for Thermal Energy Storage First Workshop*, Adana, Turkey, Apr. 16–17, 1998.
10. *Handbook of Chemistry and Physics*, 67th edition, CRC Press, 1986–87.
11. "Thermal Storage," in *2007 ASHRAE Handbook, HVAC Applications* (Ch. 34), ASHRAE, Atlanta, GA, 2007.
12. National GypsumThermalCore™ Panel, Product Literature, National Gypsum, Charlotte, NC.

CHAPTER 6

Heating Thermal Storage Systems

Introduction

As presented in the preceding chapters, a broad variety of thermal storage systems has been discussed. This variety reflects the system scale, from domestic heating and cooling to large-size power plants, relevant working temperatures, and, accordingly, different principles of operation and required materials.

It is widely accepted that heating thermal storage systems are a key element for effective thermal management in the sectors of process heat and power generation, whereas for solar thermal applications, heating thermal storage is considered indispensable. These heating thermal storage systems are diversified with respect to temperature, power level and heat-transfer media. As each application is characterized by its specific operation parameters, the understanding of a broad variety of storage designs, materials, and methods is essential.

Heat thermal energy storage (HTES) systems are used to correct the mismatch between the supply and demand of energy. This mismatch may be caused by either an intermittent energy source or significant variation in the consumption levels. With an HTES system, the provision of power is no longer coupled directly to the provision of heat. These systems can be used industrially with combined heat and power (CHP) systems to provide process steam.[1] For example, the exhaust gas of a diesel engine can be used to generate steam, which in turn is used to charge a heat storage system. When there are significant fluctuations in the facility's steam consumption—as is the case when autoclaves are filled, for example—the steam is provided by the steam generator and the HTES. An effective operation of energy converters such as absorption cooling units, Organic Rankine Cycle (ORC) processes, or Stirling engines relies on constant thermal energy supply, but, in many cases, even huge amounts of waste heat, for example, produced by a foundry, are not continuously available, with considerable fluctuations either in the power or the temperature. An HTES system can provide a constant heat flow to the energy converter, and the thermal output can be the average of the intermittent waste heat previously absorbed into the storage. Thus, the energy converter can be properly sized and operate at an optimal efficiency level. Heat thermal storage can be incorporated in a CHP system, which uses the waste heat of the turbine for thermal seawater desalination, for example, by the well-known multistage flash

---

This chapter was written by Gennady Ziskind.

(MSF) or multieffect-desalination (MED) technology.[2] The production of heat and electricity by a cogeneration system based on liquid-cooled photovoltaic cells can be significantly improved using an HTES, which not only extends the effective solar hours but also increases the range of applications, having the potential to provide heat at different power level and capacity.[3]

The role of thermal storage in solar energy utilization on any scale is obvious and follows from the fact that the heat source is inherently intermittent. It is important to stress that even if the ultimate product of the solar system is electricity, thermal storage may be preferable over electricity storage. For electricity supply one can rely on the grid to provide backup; storage of electricity (e.g., batteries) is usually expensive and is reserved for situations where the grid is not available or backup power is critical. For thermal energy, heat can be stored more easily for subsequent use,[4] at least for short and medium periods of a few hours or overnight. Thermal storage will reduce the need for conventional backup energy, and thus increase the solar share of the energy provided to the consumer, and increase the continuity and reliability of energy supply.[5]

Without energy storage, the electrical output of a solar thermal power plant would be dictated by both predictable and unpredictable variations, related to season, time, and weather. On the utility level, a fully functional thermal storage system is essential in order to mitigate the changes in solar radiation or to meet demand peaks. A distinct advantage of solar thermal power plants compared with other renewable energies, such as photovoltaics (PV) and wind, is the possibility of using relatively cheap thermal storage systems, since storing the thermal energy itself is much less expensive than storing electricity.[6] An HTES system can store energy in order to shift its delivery to a later time, or to smooth out the plant output during intermittently cloudy weather conditions. Hence, the operation of a solar thermal power plant can be extended beyond the periods of solar radiation, without the need to burn fossil fuel. Thus, times of mismatch between energy supply by the sun and energy demand can be reduced.

This chapter focuses on the HTES systems for solar-thermal power plants, and the advancements reached also benefit other applications, for example, the waste heat utilization and electricity-heat cogeneration. Solar thermal power plants produce electricity in the same way as conventional power plants, but using solar radiation as the energy input. This energy can be transformed to high-temperature steam to drive a turbine, or as heat for some other thermal engine. Of the main plant components, namely, concentrator, receiver, transport/storage media system, and power conversion device, thermal storage is both a key and the less developed component.[7] The temperature range in which the storage is performed is usually between 120 and 600°C (250–1100°F).

In order to provide a background, this chapter presents first an overview of solar-thermal power plants operating or being developed at present. Then, HTES concepts are introduced. Finally, different storage principles and corresponding materials are given, and a number of representative examples are presented.

Solar Thermal Power Plants

In solar-thermal power plants, electricity is produced using concentrated solar power (CSP). The following basic approaches are presently used: parabolic troughs, solar towers, and dish/engine systems. The type and size of the heat thermal storage depends on the plant type, as presented below.[8]

Parabolic Troughs

A parabolic trough solar collector is designed to concentrate the sun's rays via parabolic curved solar reflectors onto a heat-absorber element—a "receiver"—located in the optical focal line of the collector. The solar collectors track the sun continuously. The key components of a parabolic trough power plant are mirrors, receivers, and turbine technology. The receiver consists of a specially coated absorber tube, which is embedded in an evacuated glass envelope. The absorbed solar radiation warms up the heat-transfer fluid flowing through the absorber tube to almost 400°C (750°F). The heated heat-transfer fluid flows through a heat exchanger, in which steam is produced, which then generates power in the turbines. The output of the power plant can be between 25 megawatt (MW) and 200 MW of electricity, at its peak. Thanks to thermal storage systems, the power plant can keep working at a constant load. With high performance and low electricity production costs, the outlook for parabolic trough power plants is very good. In the recent years, systems in which water, rather than oil, flows in the tubes are being developed. The flowing water boils directly in the tubes, producing steam without a heat exchanger. For this reason, the approach is frequently termed "direct steam generation" (DSG).

Solar Towers

Solar central tower systems have a single receiver placed on top of a tower surrounded by hundreds of mirrors (heliostats), which follow the motion of the sun in the sky, and which redirect and focus the sunlight onto the receiver. The key elements of a solar tower system are the heliostats provided with a two-axis tracking system, the receiver, the steam generation system, and the heat thermal storage system. The number of heliostats will vary according to the particular receiver's thermal cycle and the heliostat design.

Dish/Engine Systems

In solar dish/engine systems, parabolic dishes capture the solar radiation and transfer it to a Stirling engine placed in the focus of the parabolic dish. A Stirling machine is a heat engine that uses external heat sources to compress and expand a working fluid, and thus to produce mechanical work/power. The dish/engine system approach is particularly suited for decentralized electricity generation, and, at present, has the highest solar-to-electricity conversion efficiency known.

It is widely assumed that one single storage technology will not meet the unique requirements of different solar power plants. Thus, highly specific design specifications are required regarding the primary heat-transfer fluid (HTF), operating pressure and temperature, power rating, and capacity.[9]

Thermal Storage Options

The principle options for using thermal storage in a solar thermal system highly depend on the daily and yearly variation of radiation and on the electricity demand profile. The main options are buffering, delivery period displacement, delivery period extension, and yearly averaging.[6]

Buffering

The goal of a buffer is to smooth out transients in the solar input caused by passing clouds, which can significantly affect operation of a solar electric generating system (SEGS) plant. The efficiency of electrical production will degrade with intermittent insolation, largely because the turbine-generator will frequently operate at partial load and in a transient mode. If regular and substantial cloudiness occurs over a short period, turbine steam conditions and/or system flow can degrade enough to force turbine trips, if there is no supplementary thermal source to "ride through" the disturbance.[6] Buffer thermal storage systems typically require small storage capacities (maximum 1 hour full load),[6] and it is quite common to design buffers as steam accumulators.[10]

A most recent example is PS10 solar thermal power plant at Sanlúcar la Mayor in southern Spain, inaugurated in 2007,[8] and its schematics and aerial view are shown in Fig. 6-1. The plant features a tower 100.5 meters (m) high and a large solar field of 624 heliostats—mobile mirrors that are faced toward the sun by a control system. Each heliostat is a mobile 120 square meter (m²) curved reflective surface mirror. The solar receiver on top of the tower produces saturated steam at 40 bar [580 pound per square inch gauge (psig)] and 250°C (482°F) and circulates it to a conventional steam turbine

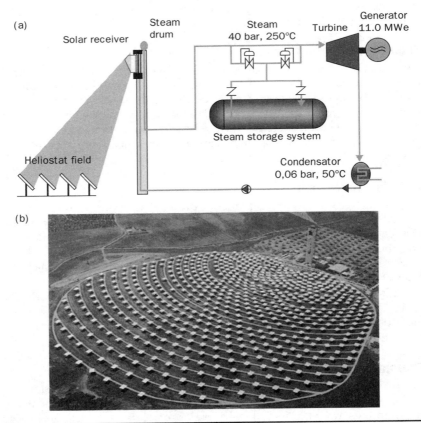

FIGURE 6-1   Schematics (a) and aerial view (b) of PS10 solar thermal power plant.[8]

that generates electricity. The nominal electrical power of the plant is 11 MW and should generate some 23 gigawatthours of electricity every year.

For cloudy periods, the plant has a saturated water thermal storage system (steam accumulator) with a thermal capacity of 20 megawatthour (MWh) [68 million British thermal unit (Btu)], with 50 MWh at 50 percent load. The system is composed of four storage tanks that are sequentially operated in relation to their charge status. During full load operation of the plant, part of the steam produced by the receiver is used to load the thermal storage system (i.e., the steam accumulator). When energy is needed to cover a transient period above the output of the plant, energy from saturated water is recovered at 20 bar (290 psig) to run the turbine at a 50 percent partial load.

Delivery Period Displacement

Delivery period displacement requires the use of a larger thermal storage capacity. The thermal storage shifts some or all of the energy collected during periods with sunshine to a later period with higher electricity demand or tariffs (electricity tariffs can be a function of hour of the day, day of the week, and the season). This type of HTES does not necessarily increase either the solar fraction or the required collection area. The typical thermal storage size for this application ranges from 3 to 6 hours of full load operation.[6]

Delivery Period Extension

The size of an HTES for delivery period extension has a typical size of 3 to 12 hours of full load,[6] and the purpose is to extend the period of power plant operation with solar energy. This TES increases the solar fraction and requires larger solar fields than a system without this type of thermal storage. Delivery period extension is the foremost objective of most of the solar power plant heat thermal storage systems used at the present.

Yearly Averaging

Yearly averaging of electricity production requires much larger HTES and solar fields. In general, these are very expensive systems and have not been given serious consideration for power plants, and are not considered here. Note, that seasonal storage is explored in other applications, for example, HVAC systems.

Definitive selection of storage capacity is site and system dependent. Therefore, detailed statistical analysis of system electrical demand and weather patterns at a given site, along with a comprehensive economic trade-off analysis in a feasibility study to select the best thermal storage capacity for a specific application is required.

Classification

Commonly, heating thermal energy storage systems are classified by storage concept (active or passive) and storage mechanism (sensible, latent, or chemical).

Storage Concepts

As discussed in Chap. 3, heating storage concepts can be classified as active or passive systems. Active storage is mainly characterized by forced convection heat transfer into the storage material. In turn, active systems are subdivided into direct and indirect systems. In a direct system, the heat-transfer fluid serves also as the storage medium, whereas in an indirect system, a second medium is used for storing the heat. In an active system, the storage medium itself circulates through a heat exchanger. This heat exchanger can be a solar receiver or a steam generator, for example.

The main characteristic of a passive system, or regenerator, is that a heat-transfer medium passes through thermal storage only for charging and discharging. The storage medium itself does not circulate. The high temperature fluid (HTF) carries energy received from the energy source to the storage medium during charging and receives energy from the thermal storage material when discharging. Passive systems are generally dual medium storage systems: the sensible heat of HTF, especially if the latter is liquid, is also used for heat storage. The thermal storage medium can be a solid, liquid, or phase-change. The main disadvantage of regenerators is that the HTF temperature decreases during discharging as the storage material cools down. Another problem is the internal heat transfer, especially for solid materials, where heat conduction is the primary heat-transfer mechanism as the heat transfer is rather low, and there is usually no direct contact between the HTF and the storage material as the heat is transferred via a heat exchanger.

Storage Media

Heating thermal storage can utilize sensible or latent heat of various substances, or heat coming from chemical reactions. As discussed in previous chapters, sensible heat is energy stored by increasing the temperature of a solid or liquid. Latent heat, on the other hand, is the means of storing energy via the heat of phase transition, usually from a solid to liquid state. A comparison of various materials is required, in terms of their storage capacity versus temperature.[11] In the chemical heat storage, it is essential that the chemical reactions involved are completely reversible.

Sensible Heat Storage

As discussed in previous chapters, thermal energy can be stored in the sensible heat (temperature change) of substances that experience a change in internal energy. The stored energy is calculated by the product of its mass, the average specific heat, and the temperature change. Besides the density and the specific heat of the storage material, other properties are important for sensible heat storage: operational temperatures, thermal conductivity and diffusivity, vapor pressure, compatibility among materials, stability, heat loss coefficient as a function of the surface areas to volume ratio, and cost.[6]

The characteristics of many candidate liquid and solid sensible heat storage materials are well-known.[6,7] For each material, the low and high temperature limits are usually considered. These limits, combined with the average mass density and heat capacity, lead to a volume-specific heat capacity in kWht (kilowatthour, thermal) per cubic meter. The approximate costs of the storage media in dollars per kilogram finally yield unit costs in $/kWht.

The average thermal (heat) conductivity given in the tables has a strong influence on the heat-transfer design and heat-transfer surface requirements of the thermal storage system, particularly for solid media (high conductivity is preferable). High volumetric heat capacity is desirable because it leads to lower storage system size, reducing external piping and structural costs. Low unit costs obviously mean lower overall costs for a given thermal storage capacity.

Liquid Media Storage

It should be noted that for temperatures exceeding 100°C (212°F), nonpressurized liquid water cannot be used as a storage medium. The use of relatively expensive pressure vessels makes this technology unattractive for higher temperatures. Other liquid storage media with higher evaporation temperatures, such as oil or molten salt can be used. Considering the temperature limitations, set by the necessity to have the material in the liquid state under the operating temperature conditions of the system, both the oils and salts are feasible in heating thermal storage, although some of them, like mineral oil, are not suitable for a solar thermal power plant.[6] The salts generally have a higher melting point, and parasitic heating is required to keep the salts liquid at night, during low insolation periods, or during plant shutdowns.[6] Silicone oil is quite expensive, though it does have environmental benefits because it is a nonhazardous material to the environment, whereas synthetic oils may be classified as hazardous materials. Nitrites in salts present potential corrosion problems, though these are probably acceptable at the temperatures required by typical systems.[6] Molten salts are used in solar power tower systems because molten salts are liquid at atmospheric pressure; provide an efficient, low-cost medium in which to store thermal energy; have operating temperatures that are compatible with today's high pressure and high-temperature steam turbines; and are nonflammable and nontoxic.[7]

Liquid media HTES typically utilize tank storage and can be designed as one tank or two tank systems.

Two-Tank Systems

An active direct two-tank system uses one tank for cold HTF coming from the steam generator and one tank for the hot HTF coming directly out of the solar receiver before it is fed to the steam generator. The advantage of this system is that cold and hot HTF are stored separately. The main disadvantage of this system is the need for a second tank. In this type of heating thermal storage system, the storage tanks are directly coupled to the HTF pressure levels (which is not necessarily a disadvantage). Early examples of relatively large-scale two-tank systems for solar electric applications are the storage systems of the SEGS I (parabolic-trough) and Solar Two (central receiver) solar thermal plants.[6]

Figure 6-2 shows a schematic diagram of SEGS I solar-thermal plant operated in the past in Daggett, CA.[6] Oil was used as both the heat-transfer fluid and storage medium. The cold and hot tank volume was 4160 cubic meter (m^3) and 4540 m^3, respectively. The thermal capacity was 120 megawatthour, thermal (MWht) or approximately 410 million Btu.

The Solar Two power plant, operated before 1999 in Barstow, CA, was a 10-MW electrical power solar tower system using a nitrate eutectic molten salt as the HTF. Molten salt was pumped from the cold storage tank through the tower receiver and then to the hot storage tank. When dictated by the operation, the hot salt was pumped from the hot storage tank through the steam generation system and then back to the cold storage tank. The system contained 1.5 million kilograms (3.3 million pounds) of nitrate salt composed of a mixture of 60% $NaNO3$ (sodium nitrate) and 40% $KNO3$ (potassium nitrate). A eutectic of nitrate salts was chosen because of the corrosivity of nitrite salts at central receiver system temperature levels.

The molten nitrate-salt thermal-storage system consisted of an 11.6-m-diameter and 7.8-m-high cold-salt storage tank, a 4.3-m-diameter and 3.4-m-high cold-salt receiver

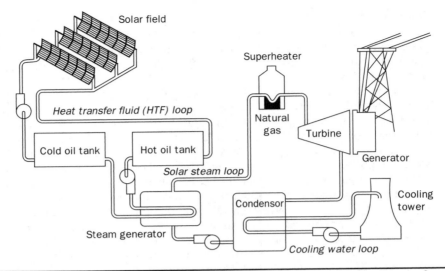

FIGURE 6-2   Schematic diagram of SEGS I solar-thermal plant with two-tank thermal storage.[6]

sump, an 11.6-m-diameter and 8.4-m high hot-salt storage tank, and a 4.3-m-diameter and 2.4-m-high hot-salt steam generator sump. The design thermal storage capacity of the Solar Two molten salt system was 105 MWht (360 million Btu), enough to run the steam turbine at full output for 3 hours. The measured gross conversion efficiency of the 12-MW electrical (10-MW net) Solar Two turbine was 33 percent. Thermal storage capacity based on the mass of salt in the tanks, accounting for (subtracting) the 3-foot heels in each tank, and with design temperatures of 1050°F hot salt and 550°F cold salt, was 114 MWht (390 million Btu). This salt melted at 220°C and was thermally stable to about 600°C.[6]

At present, the AndaSol power plant in Spain is set to convert primary solar energy into electricity using a 510,120 m² (5.5 million square-feet) parabolic trough solar field, a 7.5-hour reserve, molten-salt, two-tank, active, indirect thermal storage system and a 49.9 MW-capacity steam cycle. Without thermal storage, the solar resources in the area would only allow about 2000 annual equivalent full-load hours. The thermal storage system increases the annual equivalent full-load hours to approximately 3589 hours. The plant schematics and its aerial view are shown in Fig. 6-3.[8]

During the hours of sunlight, the parabolic trough collectors in the solar field concentrate the radiation on the absorber tubes, heating the HTF. The energy contained in this HTF can be pumped directly to the steam generator, or it can be pumped to a thermal storage system where it is stored for later use. At midday, when the sun is high in the sky, electricity can be generated and the storage system charged at the same time. The heat from the HTF is transferred to the thermal storage medium, a molten-salt fluid, which collects the heat while the salt goes from a cold tank to a hot tank, where it accumulates until the thermal storage system is completely charged (to charge storage, the salt is heated).

On discharging, the molten salt is cooled down again. During this process, the salt is liquid. The cold and hot salts are kept in separate storage tanks, so it is called a

(a) SOLAR FIELD   THERMAL STORAGE   POWER BLOCK

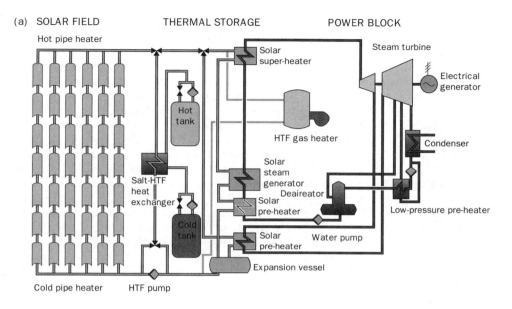

Hot pipe heater

Solar super-heater

Steam turbine

Electrical generator

Hot tank

HTF gas heater

Salt-HTF heat exchanger

Solar steam generator

Condenser

Deaireator

Solar pre-heater

Cold tank

Low-pressure pre-heater

Solar pre-heater   Water pump

Expansion vessel

Cold pipe heater   HTF pump

(b)

FIGURE 6-3 Schematics (a) and aerial view (b) of AndaSol solar thermal power plant.[8]

two-tank system. With this type of heating thermal storage, the AndaSol plant can produce solar electricity even when there is no sunshine, to dispatch the power demand at any time and ensure the rated power capacity.

As the day advances and the intensity of the solar radiation drops in the afternoon, heat is no longer sent to the storage system but employed entirely to produce electricity. After the sun sets, the solar field stops working and the storage system begins to discharge. The heat is recovered from the hot salt tank through the thermal oil to keep up electricity production during the night. To avoid deviation from scheduled production during cloudy periods, and to avoid solidification of the HTF and storage salt when electricity generation is interrupted, the plant has auxiliary gas heaters.[8]

A suggested SOLAR TRES power plant in Spain, shown schematically in Fig. 6-4, is a further development of the solar-tower approach used in Solar Two.[8] It includes a 2480-heliostat field with 120 m² glass metal heliostats, a 120 MW (thermal) high-thermal-efficiency cylindrical receiver system, and an active, direct thermal storage system (15 hours, 647 MWh, 6250 tons liquid nitrate salt) with insulated tank immersion heaters. The high-temperature liquid salt at 565°C in stationary storage drops only 1 to 2°C per day. The cold salt is stored at 45°C above its melting point (240°C), providing a substantial margin for design. Advanced pump designs will pump salt directly from the storage tanks, eliminating the need for pump sumps, and high-temperature multistage vertical turbine pumps mounted on top of the thermal storage tanks, using a long-shafted pump with salt-lubricated bearings. This pump arrangement also eliminates the sump, level control valve, and potential overflow of the pump sump vessels. A 43-MW steam generator system will have a forced recirculation steam drum, and this innovative design places components in the receiver tower structure at a height above the salt storage tanks that allows the molten-salt system to drain back into the

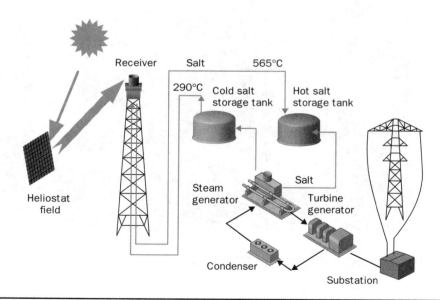

FIGURE 6-4 Proposed SOLAR TRES power plant in Spain.[8]

tanks, providing a passive fail-safe design. This simplified design improves plant availability and reduces operation and maintenance costs.[8]

Single Tank Systems

A single tank system is similar in concept to a domestic solar water heater. The single-tank system reduces storage volume and cost by eliminating a second tank.[6] However, in a single-tank system it is more challenging to separate the hot and cold HTF; however, as discussed in previous chapters, liquid media maintain natural thermal stratification because of density differences between hot and cold fluid. Thus, the HTF naturally stratifies in the tank, from coolest layers at the bottom to the warmest layers at the top. These systems are called stratified thermal storage. Maintaining the thermal stratification requires the hot fluid to be supplied to the upper part of a storage system during charging and the cold fluid to be extracted from the bottom part during discharging, or the use of another mechanism to ensure that the fluid enters the storage at the appropriate level in accordance with its temperature (density) in order to avoid mixing. This can be done by some stratification devices (floating entry, mantle heat exchange, etc.). Depending on the cost of the storage fluid, the stratified thermal storage can result in a substantially lower cost storage system. However, the stratified storage system must maintain the thermocline zone in the tank, so that it does not expand to occupy the entire tank.

Among the systems built in the past that used a single tank stratified thermal storage system were the irrigation pump in Coolidge, AZ [tank volume 114 cubic meters (m^3), storage capacity 3 MWht] and IEA-SSPS in Almeria, Spain (tank volume 200 m^3, storage capacity 5 MWht).[6] Both systems were based on the parabolic-through technology, and were active direct, using oil as both the HTF and storage medium.

Filling the storage tank with a second solid storage material (rock, iron, sand, etc.) can help to achieve the stratification. Among the systems that used a single tank/dual medium in the past is the IEA-SSPS in Almeria, Spain (parabolic-trough, oil/cast iron, tank volume 100 m^3, storage capacity 4 MWht) and Solar One in Barstow, CA (central receiver, oil/sand/rock, tank volume 3460 m^3, storage capacity 182 MWht).[6]

Actually, an additional advantage of a stratified thermal storage system is that most of the storage fluid can be replaced with a low-cost filler material. Sandia National Laboratories has demonstrated and tested a 2.5-MWh, packed bed, stratified thermal storage system with binary molten-salt fluid, and quartzite rock and sand for the filler material.[12]

Solid Media Storage

Liquid storage media with higher evaporation temperatures like oil or molten salt could have some disadvantages, for example, degradation or high freezing points. Thus, an attractive option regarding investment and maintenance costs is sometimes the application of solid sensible heat storage media.

Actually, the dual system described above might be alternatively included in the "solid media" category, as it may be seen as a solid packed bed, requiring a fluid to exchange heat. The lower the heat capacity of the fluid, the more solid the storage. When the heat-transfer fluid is a liquid, its capacity is significant, and the storage system is essentially a dual one, whereas when the fluid heat capacity is very low (e.g., when using air) the solid is the only storage material. Packed beds favor thermal stratification so that stored energy can easily be extracted from the warmer strata, and cold fluid can be taken from the colder strata and fed into the collector field. However, pressure drop and, thus, parasitic energy consumption may be high in a dual system, which has to be considered in the HTES design.[6]

Among the materials considered for thermal storage applications, the sand-rock-oil combination is not suitable for the temperatures above 300°C. Reinforced concrete and salt have low cost and acceptable heat capacity but very low thermal conductivity. Silica and magnesia fire bricks, usually identified with high temperature thermal storage, offer no advantages over concrete and salt at these lower temperatures. Cast steel is too expensive, but cast iron offers a very high heat capacity and thermal conductivity at moderate cost.

A solid media concrete storage based on in-tube HTF is suggested by the German Aerospace Center (DLR) in Stuttgart[13] and is essentially a dual medium indirect storage system with regenerative heat transfer, with modular and scalable design from 500 kWh to 1000 MWh, and flexible to a large number of sites and construction materials. The tube register consists of 132 tubes with a length of 9 m and an outer diameter of 18 millimeter. The tubes are arranged in a triangular pitch. Collectors consisting of flat plates and semi-shells are placed at the end of the tubes. The tube register is prefabricated in the workshop, transported on-site, and lifted onto the previously prepared foundation by crane. As preparatory work, a large number of material tests with various kinds of cements, aggregates, and admixtures had been carried out. Finally, a special concrete mixture using polyethylene fibers was used for the thermal storage.[13]

Concrete storage has so far been designed for parabolic trough solar thermal power plants of the AndaSol-type, using thermal oil as heat-transfer fluid. For this 50 MW electrical plant a concrete storage with an overall capacity of approximately 1100 MWh will be built up modularly from 252 basic storage modules with about 400 tons of concrete each. These basic modules will be connected in series and parallel and packed together into four insulated storage units of 63 modules. A ground surface of about 300 × 100 m is required.[13]

Latent Heat Storage

As discussed in previous chapters, thermal energy can be stored nearly isothermally in some substances as the latent heat of phase change, that is, the heat of fusion (solid-liquid transition), the heat of vaporization (liquid-vapor), or the heat of solid-solid crystalline phase transformation. Note, however, that the name "phase change materials" (PCM) often implies that the solid-liquid phase change takes place.[14,15]

Because the latent heat of fusion between the liquid and solid states of materials is rather high compared to the sensible heat, storage systems utilizing PCMs can be reduced in size compared to single-phase sensible heating systems. However, heat-transfer design and media selection are more difficult, which may be related to the lack of materials with the required melting temperature and the fact that the performance of the materials can degrade after a moderate number of freeze-melt cycles. Different salts are suggested for the purpose of heat storage.[16] It is important to note that organic substances, like paraffins, are typically not suitable for heat thermal storage since their melting temperatures are usually below 100°C.

It is widely accepted that in order to take full advantage of PCM storage, several modules with different materials and different melting points should be connected in series.[6,17–19] The reason for needed multiple PCM cascade storage is that for a single-stage storage during discharging, the HTF temperature, in the biggest part of the storage module, is higher than the melting temperature of the salt, which means that a major portion of the salt would not freeze during discharging, and the high latent portion of the stored heat cannot be extracted from the storage.[6] Consequently, the utilization factor of the system would be relatively low. Accordingly, the melting temperatures of the PCM should be cascaded from the lower operating temperature upward to the maximum operating temperature, which is 390°C for conventional parabolic trough power plants, with synthetic heat transfer oil to match the different characteristics of the sensible heat-transfer medium and the latent storage media. The concept may also be applicable for DSG technology (DSG; sometimes also referred to as Direct Solar Steam, DiSS) for temperatures up to 500°C where the use of cascades is only needed for the preheating and superheating of the water/steam, whereas the evaporation excellently fits the latent heat storage characteristics.

In order to take advantage of the latent heat proper, small temperature differences between the HTF and the PCM are desirable. Several approaches to reduce the thermal resistance between the HTF and the PCM has been suggested, for example, packing the PCM within a high thermal conductivity solid matrix, mixing high thermal conductivity particles into the PCM, increasing the area for heat transfer by utilizing extended surfaces, and encapsulating the PCM within the HTF.

Among the recent developments are innovations with PCM-concrete and PCM-graphite composite heat storage, for use in DSG, where boiling water is used instead of single-phase oil in the parabolic-trough solar collectors. Although, the two-phase heat-transfer fluid poses a major challenge both for the plant design as a whole and for its thermal storage system. A three-part storage system is proposed where a PCM storage will be deployed for the two-phase evaporation, whereas concrete storage will be used for storing sensible heat, that is, for preheating of water and superheating of steam.[20]

The system consists of a concrete preheater unit, a PCM evaporator unit, and a concrete superheater unit. During discharging, feedwater enters the preheater and is heated close to the boiling curve, then the water is circulated through the PCM storage, where part of the water evaporates. The steam is separated from the water in the steam drum, and the steam is superheated in the concrete unit, while the remaining water is recirculated through the PCM storage.[20]

The DISTOR project, sponsored by the European Union,[8] focuses on the development of PCM for more efficient energy storage, under three main themes: developing innovative storage media based on the microencapsulation of PCM in a matrix of expanded graphite; selecting the most efficient storage configuration for high efficiency internal charging/discharging heat transfer; and demonstrating the feasibility of the developed material and storage design through in situ testing of the storage module on a 100-kW test prototype under real field conditions.

The innovative approach to increasing storage efficiency taken by DISTOR is the micro-encapsulation of PCM in a matrix material with high porosity and superior thermal conductivity. Limiting the size of PCM clusters in the matrix, the thermal conductivity of the total composite will be controlled by the matrix of expanded graphite. The feasibility of the micro-encapsulation concept has already been demonstrated with paraffin PCM for applications in temperatures below 100°C. For higher temperature ranges, salt materials have to be used as storage media.

Alternatively, it has been suggested to incorporate devices such as heat pipes to enhance the heat transfer between the HTF and the PCM. It is well-known that heat pipes have very high effective thermal conductivities, can be tuned to operate passively in specific temperature ranges, and can be fabricated in a variety of shapes.[21]

Chemical Thermal Storage

A third storage mechanism is by means of chemical reactions.[6] For this type of heating thermal storage, it is necessary that the chemical reactions involved are completely reversible. The heat produced by the solar receiver is used to excite an endothermic chemical reaction. If this reaction is completely reversible, the heat can be recovered completely by the reversed reaction. Often catalysts are necessary to release the heat, which is advantageous as the reaction can then be controlled by the catalyst.

Commonly cited advantages of HTES in a reversible thermochemical reaction (RTR) are high storage energy densities, indefinitely long storage duration at near ambient temperature, and heat-pumping capability. Drawbacks may include the following: complexity; uncertainties in the thermodynamic properties of the reaction components and of the reaction kinetics under the wide range of operating conditions; high cost; toxicity; and flammability.

Although RTR have several advantages concerning their thermodynamic characteristics, development is at a very early stage. To date, no viable prototype plant has been built, although various reactions are being investigated for chemical heat storage.[7] It is argued that the reactions involving metal oxides/metals and ammonia are two promising methods for RTR.[7]

It appears that conversion of concentrated solar high-temperature heat to chemical fuels has the advantage of producing long-term storable and transportable energy carriers from solar energy.[22,23] In particular, solar energy could be used for hydrogen production. Among other ideas, hydrogen production from water using the sulfur-iodine (S-I) thermochemical cycle, powered by combined solar and fossil heat sources, has been investigated.[24] It was conceived in order to operate the chemical process continuously: a solar concentrator plant with a large-scale heat storage supplies thermal load for services at medium temperatures (<550°C), while a fossil fuel furnace provides heat load at higher temperatures. It is argued that hydrogen produced with electricity from solar thermal plants will be one major option for clean fuel alternatives in the future.[2]

References

1. Christ, M., M. Franz, M. Haas, W. Eden, and W.-D. Steinmann, "Hidden Energies—Industrial Uses for Thermal Storage," *DLR Nachrichten*, 120: 56–57, 2008.
2. Quaschning, V., and F. Trieb, "Solar thermal power plants for hydrogen production," *HYPOTHESIS IV Symposium*, Stralsund, Germany, Sept. 9–14, 2001, 198–202.
3. Kribus, A., R. Tamme, and G. Ziskind, Development of an advanced solar cogeneration system with concentrating photovoltaics and phase change material thermal storage, Proposal for project funding in German-Israeli cooperation in energy research, 2005.
4. Dincer, I., and M. A. Rosen, "Energy Storage Systems: Systems and Applications," in *Thermal Energy Storage* (Ch. 2), eds. I. Dincer and M .A. Rosen, John Wiley & Sons, 2002.

5. Dincer, I., and M. A. Rosen, "Thermal Energy Storage (TES) Methods," in *Thermal Energy Storage* (Ch. 3), eds. I. Dincer and M. A. Rosen, John Wiley & Sons, 2002.

6. Pilkington Solar International GmbH, Survey of Thermal Storage for Parabolic Trough Power Plants, National Renewable Energy Laboratory, Sept. 2000, NREL/SR-550-27925.

7. Gil, A., M. Medrano, I. Martorell, A. Lázaro, P. Dolado, B. Zalba, and L. F. Cabeza, "State of the Art on High Temperature Thermal Energy Storage for Power Generation. Part 1—Concepts, Materials and Modellization," *Renewable and Sustainable Energy Reviews*, 14: 31–55, 2010.

8. European Commission, Directorate-General for Energy and Transport, Concentrating solar power—from research to implementation, Office for Official Publications of the European Communities, Luxembourg, 2007, ISBN 978-92-79-05355-9.

9. D. Laing, Solar thermal energy storage technologies, Energy Forum, 10,000 Solar GIGAWATTS, Hannover, April 23, 2008.

10. Steinmann, W.-D., and M. Eck, "Buffer storage for direct steam generation," *Solar Energy*, 80: 1277–1282, 2006.

11. D. Laing, "Concrete Storage Development for Parabolic Trough Power Plants," *Parabolic Trough Technology Workshop*, Mar. 08, 2007, Golden, CO.

12. Pacheco, J. E., S. K. Showalter, and W. J. Kolb, Development of a Molten-Salt Thermocline Thermal Storage System for Parabolic Trough Plants, *Forum 2001: Solar Energy: The Power to Choose*, April 24, 2001.

13. Laing, D., D. Lehmann, and C. Bahl, "Concrete storage for solar thermal power plants and industrial process heat, IRES III 2008," *3rd International Renewable Energy Storage Conference*, Berlin, Nov. 24–25, 2008.

14. Kenisarin, M., and K. Mahkamov, Solar Energy Storage Using Phase Change Materials, *Renewable and Sustainable Energy Reviews*, 11: 1913–1965, 2007.

15. Kenisarin, M. M., High-temperature phase change materials for thermal energy storage, *Renewable and Sustainable Energy Reviews*, 14: 955–970, 2010.

16. Bauer, T., R. Tamme, M. Christ, and O. Öttinger, "PCM-Graphite Composites for High Temperature Thermal Energy Storage," *The Tenth International Conference on Thermal Energy Storage*, Atlantic City, 31 May—2 June 2006.

17. "Phase-change thermal energy storage," *Final Subcontract Report on the Symposium held*, Oct. 19–20, 1988, Helendale, CA, Luz International Ltd., Solar Energy Research Institute, CBY Associates, Inc., November 1989.

18. Michels, H. and E. Hahne, "Cascaded Latent Heat Storage for Solar Thermal Power Stations, EuroSun.96," *Proceedings of 10th Int. Solar Forum*, Freiburg, Germany, 1996.

19. Michels, H., and R. Pitz-Paal, Cascaded Latent Heat Storage for Parabolic Trough Solar Power Plants, *Solar Energy*, 81: 829–837, 2007.

20. Laing, D., C. Bahl, T. Bauer, D. Lehmann, and W.-D. Steinmann, Thermal Energy Storage for Direct Steam Generation, *Solar Energy*, 2010, in press, doi:10.1016/j.solener.2010.08.015

21. Shabgard, H., T. L. Bergman, N. Sharifi, and A. Faghri, High Temperature Latent Heat Thermal Energy Storage Using Heat Pipes, *International Journal of Heat and Mass Transfer*, 53: 2979–2988, 2010.

22. Gokon, N., S. Inuta, S. Yamashita, T. Hatamachi, and T. Kodamac, Double-Walled Reformer Tubes Using High-Temperature Thermal Storage of Molten-Salt/MgO

Composite for Solar Cavity-Type Reformer, *International Journal of Hydrogen Energy,* 34: 7143–7154, 2009.

23. Ma, Q., L. Luo, R. Z. Wang, and G. Sauce, A Review on Transportation of Heat Energy Over Long Distance: Exploratory Development, *Renewable and Sustainable Energy Reviews*, 13: 1532–1540, 2009.

24. Giaconia, A., R. Grena, M. Lanchi, R. Liberatore, and P. Tarquini, Hydrogen/ methanol production by sulfur–iodine thermochemical cycle powered by combined solar/fossil energy, *International Journal of Hydrogen Energy,* 32: 469–481, 2007.

CHAPTER 7

Thermal Storage System Sizing

Introduction

This chapter discusses and provides examples of thermal storage system sizing for chilled water (CHW), hot water (HW), and ice storage systems, which as previously noted are among the most common type of thermal storage systems in use today. The key input for any thermal storage system is the total thermal load over the period evaluated, using the load profile for the period with the greatest load. The load profile means the load per each unit of time. If, for example, the loads are calculated hourly and the period under evaluation is a day, then the load profile is the load for each hour of the day. The total of all those hourly loads for a 24-hour period is the peak-day load. The highest load in a daily load profile is the peak hourly load. If the period under consideration is a 1-day cycle, then the daily production capacity must equal the daily load (plus any storage losses). Each hour may have more or less storage use or capacity production as long as the total daily production meets the total daily demand and as long as the thermal storage capacity is great enough to handle the sum of the hourly differences.

There are a number of methods or tools available to estimate the peak thermal load profile. Thermal storage systems are sized on the peak thermal load profile (e.g., the 0.4 percent ASHRAE weather data, facility peak-day cooling or heating load profile). The production capacity and storage capacity must meet that maximum peak-day load, even though most of the time there is more thermal storage capacity than is required.

Given the peak-day heating or cooling load profile, the thermal storage capacity in ton-hours (or other appropriate units) as well as the total minimum equipment generation capacity required can be calculated for either a full storage or partial storage operating strategy. Full storage strategies are usually used where demand reduction for a certain period of the day is the prime motivation for installing thermal storage. Full storage strategy, for example, can provide for maximum demand reduction for a given number of hours per day, since during that high-demand cost period, all generation equipment is turned off, with the facility load provided solely from thermal storage. Because, during non-on-peak hours, the production equipment must meet the instantaneous load in addition to storing thermal capacity for the period the equipment is off, this full storage strategy requires more refrigeration (or heating) equipment than partial storage systems. The amount of thermal storage needed depends on the loads during the period served by thermal storage and is also dependant, in large part, on the amount of time when the loads are served by storage.

Partial storage strategy (using a load-leveling approach) allows for the minimum total thermal production equipment capacity required to meet the peak thermal load, and production equipment is usually much smaller than for conventional systems without thermal storage. If the thermal storage system is based on a daily cycle, then the minimum equipment size is the total peak-day hourly loads divided by 24 hours. Thermal storage provides an averaging effect by storing or using capacity depending on whether the instantaneous load is greater or less than the thermal production level.

Once the thermal storage capacity has been calculated based on the peak load profile and operating strategy, the actual physical size of the thermal storage tank will depend on the type of thermal storage system, and, for sensible thermal storage systems, the size of the delta-T (i.e., the size of the temperature difference between the system supply and return temperatures). In addition, there is always some lost or ineffective storage volume such as the thermocline between supply and return.

As discussed in previous chapters, the size of the delta-T is a critical factor in the physical size of sensible thermal storage systems such as CHW and HW thermal storage, but the size of the delta-T is not a critical factor with latent thermal storage systems such as ice thermal storage systems. Because of this, the operation of coils, controls, and other parts of the distribution system is critical with sensible storage systems but less so with latent thermal storage.

CHW thermal storage systems are typically sized to either (1) reduce electric on-peak demand associated with cooling production to the maximum extent possible or (2) reduce cooling plant equipment cost and associated infrastructure to the maximum extent possible. However, sometimes, thermal storage systems are also sized to reduce electric demand over a longer period of time in order to forgo costly infrastructure improvements. In contrast, HW systems may operate during the day or at night storing HW for later use in the mornings or evenings when generation equipment and the heat source are off-line.

Ice thermal storage systems follow a similar operating strategy as CHW, with respect to full storage versus partial storage operating strategies. However, as discussed in Chap. 5, internal melt ice thermal storage systems have varying supply temperatures and charge and discharge rates depending on the state of the ice charge (i.e., how much ice is built up or melted on the heat exchanger surface). Therefore, the size of the required discharge rate and when that discharge rate is needed (e.g., at the beginning of the ice thermal storage discharge cycle versus near the end of the ice thermal storage discharge cycle) can affect the physical size of an ice thermal storage system.

Determining or Estimating Facility Loads

Determining or estimating the existing and future (if facility expansion is planned) thermal load profiles is critical to the thermal storage sizing process. As discussed in Chap. 2, facility heating, cooling, and electric loads are dependent upon a large number of factors including the design and construction of a building (or buildings for a campus environment) and its or their occupancy type and schedule, for example. Additionally, facility loads vary over time (daily and seasonally); and each type of facility or building load, whether it is heating, cooling, or electrical, when graphed, has its own typical profile shape of load magnitude versus time. Note that facility loads are often interrelated. For example, the electric load is, in part, typically a function of the HVAC load, with sometimes about 40 percent of a building's electric demand used to provide facility cooling (the AC in HVAC) in warmer climates. In addition, electric loads due to lights and equipment contribute to the facility cooling load when in the cooling mode or reduce required heat input in the heating mode. It should also be noted that a large central plant serving many buildings sees a load profile that is different from the profile of the individual buildings, as there is an aggregating effect between buildings with an additional storage or flywheel effect in the mass of the return mains from the buildings.

There are a number of different methods available to produce or develop those necessary load profiles. For existing facilities and systems without loads being added, the best method is often to use installed energy management systems and examine the record of actual existing loads and energy use. Actual readings, in theory, should be more accurate than other forms of load estimating, provided that the facility meters are properly installed and calibrated and provided that there are no changes in the loads served and that the year under examination is similar to a peak heating or cooling load year (and providing that actual loads do not exceed the operating capacity of the cooling or heating plant, an occurrence that can be identified by the drifting of supply and return temperatures away from the design or set-point values).

If the necessary British thermal unit (Btu) (flow and temperature) and power meters and energy management or building control systems are not available, the next best method is to install temporary meters and data loggers to record actual load data or energy use. If possible, the load data should be recorded during the appropriate load conditions (i.e., during peak cooling loads for cooling storage and during peak heating loads for heating storages). It is important to realize that many factors affect peak loads. For example, often a college campus shows peak cooling loads in the fall when outside-air temperatures may be lower but campus use is higher than those during the warmer summer periods when campus use is less. As weather can vary from season to season in some climates, care must be taken to ensure, if possible, that the peak-day profile is captured or to make necessary adjustments. Adjustments can be made, for example, based on the peak outside-air temperature during the data gathering versus the peak design outside-air temperature, considering to the extent that the facility load is temperature driven or internal load driven.

If actual facility load data are not available either from existing installed or from temporary meters or recorders, other methods or a combination of methods can be used to estimate the facility loads. Utility use data can provide a basis for estimating the loads. For example, electrical data are usually available for existing buildings including use and demand, including peak demand, which gives good insight into the overall load patterns for the buildings. An average building with electric-driven cooling equipment will usually follow a predictable pattern of when peak cooling load and peak electrical demand occur. Naturally, buildings with high electric use for purposes other than lighting, fan power, cooling, and pumps (e.g., electric heat) may not fit this pattern. For heating storage, given the monthly natural gas (NG) usage and knowledge of existing facility NG-consuming devices, one can estimate the monthly NG used to produce just the thermal load (e.g., by deducting from the total measured NG usage any estimated nonthermal load NG use such as that in kitchen, laboratory, laundry, etc.). The total monthly thermal NG usage (in units of energy) is multiplied by the facility average boiler efficiency to obtain the estimated monthly thermal load. Given an idea of the shape of the facility heating load profile, and knowing the facility operating schedule, one can generate a 24-hour heating load profile that when totaled for the operating days in the month approximately equals the monthly NG consumption. Space heating loads in buildings in cooler climates used less than 24 hours a day are significantly dependant on warm-up loads. Most of the heat needed is often used in a short period prior to occupancy (the warm-up period), and later, after the building mass is up to temperature, and as lights and occupants contribute heat to the spaces, the heating loads are usually much less than the warm-up load. Only some of the heating load is dependent on outside-air temperatures.

As a rough method of calibrating utility-based load estimates, or even as a way to establish the peak loads themselves, some idea of peak thermal loads can be determined from installed thermal equipment capacity in conjunction with facility staff interviews (e.g., the staff may state that the equipment is undersized, or, conversely, that the full capacity of the equipment is never needed to meet the thermal load). In many cases, the thermal capacity of the installed air-handling equipment is far more than the actual load served. A major exception is where loads have been added after the original design. One of the more common situations where this occurs is where significant electronic equipment have been added.

Loads can also be estimated based on facility floor area and rule-of-thumb load factors for the specific use and construction. For example, knowing or calculating that the cooling load is approximately 400 square feet per ton for a typical classroom and given the facility's gross floor area (in square feet in this case), one can calculate the estimated peak cooling load. Construction, climate, and occupancy use differ so much that this type of load estimating should only be used as an order-of-magnitude check against some more rigorous calculation. The shape of the load profile can also be estimated given knowledge of the facility's building type, occupancy schedule, climate, and peak and base thermal loads. Given the peak hourly thermal load and the shape of the load profile, one can estimate the peak-day thermal load profile. This floor area and load factor method may be the only tool available to estimate future facility thermal loads, and it is often best when compared with actual measured existing facility load factors for similar buildings with similar uses in similar climates. An additional caveat with this method is that, with respect to building design and loads, the building industry is presently undergoing a major transformation toward net zero energy buildings for new construction, so the rule of thumb load factors used in the recent past or from existing facility data may not be accurate predictors of future load factors (i.e., future buildings are expected to be a lot more energy efficient and may actually consume or require no energy on an annual basis even though the building still experiences a thermal load).

Often the above methods are used in conjunction with energy use calculation software that, for each hour of the year, calculates the heating and cooling thermal load. Based on user model configurations and inputs, energy software can also calculate the building and plant equipment capacity needed to meet the thermal load, the required electricity, and fuel energy needed to operate facility or building and plant equipment, as well as calculate the monthly and annual energy costs. To successfully use a computer model (i.e., energy software) to estimate plant operation and energy costs or savings, the model should be calibrated to match existing facility or building energy consumption (unless the facility is brand new). It is possible to simplify the facility or building model itself, especially when numerous buildings or numerous zones exist by creating a "dummy" building that mimics the energy consumption of multiple buildings. Once one has a calibrated computer model using energy software (which is available from a number of vendors), the computer model can be used: to estimate peak future loads and load profiles; to model various alternatives including, for example, full and partial thermal storage; and to estimate energy use and cost for each alternative.

Given an understanding of the existing peak thermal loads and load profiles by any of the above methods, the thermal storage system can be sized. As noted, the size of the thermal storage system depends not only upon the thermal load profile but also upon the operating strategy employed, full storage or partial storage, the applicable utility tariff, as well as the type of thermal storage system used. A CHW thermal storage system, for example, is sized somewhat differently from an ice storage system. In addition, for example, CHW thermal storage system sizing depends upon the type of thermal storage system (e.g., multiple tank system versus stratified CHW thermal storage).

In any case, all daily thermal storage sizing begins with the peak-day load profile, which establishes the peak total daily demand, which, as noted, must be met by the thermal generation equipment operating in conjunction with the thermal storage system.

Stratified CHW Thermal Storage

The following two examples show two different methods to size a stratified CHW thermal storage system based on daily load shifting. The principles discussed for sizing stratified thermal storage systems are also often applicable to other types of thermal storage systems. The CHW thermal storage system size includes both the thermal capacity and physical size of the thermal storage tank itself and the capacity of the chillers needed to meet the load and to charge the thermal storage system. As discussed, in order to determine the size of a thermal storage system, the peak-day 24-hour cooling load profile must be available either from actual facility data or from engineering cooling load estimates. In the case of CHW thermal storage system sizing, the peak-day 24-hour cooling load profile is the day with the maximum total ton-hours, which is often (but not always) also the day with the highest instantaneous peak cooling load.

For CHW thermal storage, as previously discussed, two different sizing options are often considered: "full storage" and "partial storage." With full storage, the thermal storage system is sized so that on the peak day, with the exception of CHW distribution pumps, the chiller plant can be completely turned off for a certain period of the day, usually the utility company's on-peak rate period. With partial thermal storage (in the load-leveling configuration), the thermal storage system is sized such that, on the peak cooling day, the smallest chiller plant possible can meet the peak cooling load by operating at a constant rate over a 24-hour period in conjunction with the thermal storage system. Therefore, with partial storage, the chiller plant is operated during the on-peak period but at a lower rate than with a conventional chiller plant without thermal storage.

A third sometimes used approach is the "maximum shift" partial storage system. This often applies where the owner requires a certain amount of chiller plant redundancy in installed chiller plant capacity relative to peak loads, for example, an "N+1" design where "N" chillers would be adequate to meet the peak load, but "N+1" chillers are installed so that even if one chiller were to be unavailable, there would still be adequate chiller capacity to meet the peak load. In the maximum shift partial thermal storage approach, the storage would be sized to be able to accept the full amount of excess off-peak chiller capacity (above the off-peak loads), which can be produced using all installed chillers, including any redundant chiller capacity. In this case, the chiller plant size is the same as that used for a load-leveling partial storage system; but the thermal storage is larger, and the amount of load management is larger. Capital cost is also larger but not as much as would be the case for a full storage system where the installed chiller plant capacity would also need to be increased.

Full Storage System Sizing

To determine the required full storage thermal storage capacity and corresponding required total chiller plant capacity, one can construct a simple hour-by-hour spreadsheet model as shown in Table 7-1. The total daily ton-hours is the sum of the day's 24-hourly average cooling loads (i.e., the area under the 24-hour cooling load profile), and in the example shown in Table 7-1, this value is equal to 44,420 ton-hours.

The total 24-hour chiller plant production output in ton-hours must equal the facility's total daily ton-hour demand in ton-hours. In the case of full storage systems, the facility or building plant chillers will be off during the utility on-peak period, which in this example is a 4-hour period from 1:00 to 5:00 PM (however, the time period could just as easily be 6 hours or some other number of hours). Therefore, the minimum chiller plant capacity is the total load served divided by the number of hours the chiller can operate. In this example, the cooling load is calculated as 2221 tons (44,420 ton-hours served divided by 20 hours of chiller operation). This chiller output is entered into the chiller output column in the spreadsheet sizing model in each of the hours the chiller is allowed to operate, as shown in Table 7-1. Note that minor chiller output adjustments can be made to equalize the total daily ton-hours as necessary. If actual chiller plant capacity is or will be larger than the calculated minimum chiller plant capacity required, one can use the capacity of the larger chiller plant in the spreadsheet model, which can allow the calculated size of the proposed thermal storage tank to be somewhat smaller than in the case with the minimum-sized chiller plant capacity (see discussion on the following pages).

The last column in Table 7-1 calculates the total thermal storage tank charge for each hour, as explained below, and the required thermal storage capacity is simply the maximum storage charge value calculated in the thermal storage tank inventory charge column, which in this example is 12,641 ton-hours.

The first step in calculating the required thermal storage tank size is to identify the hour at which the thermal storage tank runs out of capacity and to enter a zero value in that cell. The thermal storage tank inventory charge column shows the amount of capacity remaining after the capacity use and production for that hour. Zero means that, at the end of the hour listed, the thermal storage capacity is depleted. Often, that time is the end of the period where the chillers are turned off, but, if the facility load at that time is greater than the chiller capacity, the capacity should be zero at a later period when the chiller capacity will exceed the cooling load. When the chiller capacity is greater than the load, thermal capacity is stored (charge). If the facility load is greater than the chiller output, then thermal capacity is taken from thermal storage (discharge). Observe that the thermal storage tank, therefore, by design, becomes fully discharged after the chillers are off and before the hour at which there is surplus chiller production.

Time of day	Facility hourly CHW load (tons)	Chilled water thermal storage system	
		Hourly chiller output (tons)	Thermal storage charge inventory (ton-hours)
12:00 AM	1,090	2,221	3,359
1:00 AM	1,019	2,221	4,561
2:00 AM	959	2,221	5,823
3:00 AM	975	2,221	7,069
4:00 AM	969	2,221	8,321
5:00 AM	968	2,221	9,574
6:00 AM	1,274	2,221	10,521
7:00 AM	1,358	2,221	11,384
8:00 AM	1,457	2,221	12,148
9:00 AM	1,773	2,221	12,596
10:00 AM	2,176	2,221	12,641
11:00 AM	2,506	2,221	12,356
12:00 PM	2,668	2,221	11,909
1:00 PM	2,831	0	9,078
2:00 PM	2,900	0	6,178
3:00 PM	2,900	0	3,278
4:00 PM	2,784	0	494
5:00 PM	2,588	2,221	127
6:00 PM	2,348	2,221	**0**
7:00 PM	2,077	2,221	144
8:00 PM	1,945	2,221	420
9:00 PM	1,836	2,221	805
10:00 PM	1,711	2,221	1,315
11:00 PM	1,308	2,221	2,228
Total	**44,420**	**44,420**	
Thermal storage size (ton-hours)			**12,641**

TABLE 7-1   Peak-Day Full Thermal Storage Sizing Calculations

So as shown in Table 7-1 full storage example, the facility cooling load falls below the chiller plant output at 7:00 PM, and therefore, the zero value is entered for thermal storage charge at 6:00 PM. The formula for thermal storage tank capacity at 7:00 PM and for every other hour in the day is to take the previous hour's thermal storage tank charge, which remains at the end of that hour, and add to that charge value the difference between chiller plant output and facility cooling load for the current hour. When the difference between chiller plant output and facility cooling load is positive, the chiller system is "charging" the thermal storage tank. Conversely, when the difference between chiller plant output and facility cooling load is negative, the thermal storage tank is "discharging" to help serve the facility's cooling load demand. In this example, at the end of the 7:00 PM hour, the thermal storage tank charge is 144 ton-hours, equal to the remaining 6:00 PM thermal storage tank charge of 0 ton-hours plus 2221 tons of chiller production minus 2077 tons of facility load. Similarly, at the end of 8:00 PM, the thermal storage tank charge is 420 ton-hours, equal to the end of 7:00 PM thermal storage charge of 144 ton-hours plus the 8:00 PM 2221 chiller production tons minus 1945 tons of facility load. The midnight hour thermal storage tank charge at the top of the column references the 11:00 PM thermal storage tank charge value at the bottom of the column (imagine two peak days in a row).

To recap, for the example shown in Table 7-1, the CHW thermal storage tank is completely discharged of CHW by the end of the 6:00 PM hour. Starting at 5:00 PM (which is the end of the utility on-peak period in this example), the chiller plant is operated, and, on a peak day, continues to operate to serve the facility cooling loads and to charge the thermal storage tank until the start of the utility on-peak demand period at 1:00 PM on the following day. Further, in this example, at 11:00 AM, the facility cooling load rises above the chiller plant output and the thermal storage tank begins the discharge mode. Therefore, the maximum thermal storage charge occurs at the end of the 10:00 AM hour, which is 12,641 ton-hours in this case. Per design, the chillers are completely shut down for the 4-hour on-peak period of the day, and the CHW thermal storage tank is discharged to deliver required CHW to serve the facility on-peak cooling loads. When the thermal storage tank is fully discharged, the daily process starts again. Figures 7-1 and 7-2 show this example's full thermal storage system operations graphically.

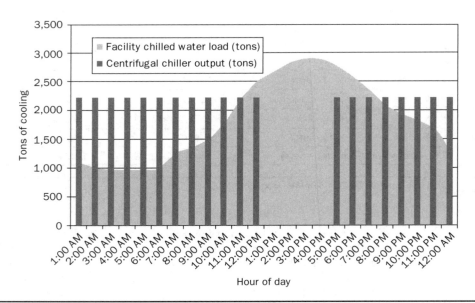

FIGURE 7-1   Full thermal storage operation: facility peak-day cooling load vs. chiller output.

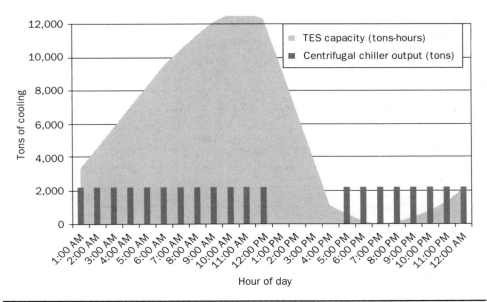

FIGURE 7-2   Full thermal storage capacity and chiller output.

Depending on how much of a safety factor is in the peak-day 24-hour cooling load profile, it is probably prudent to add some additional ton-hour capacity to the calculated minimum thermal storage capacity required and to add some additional tonnage to the calculated minimum chiller plant capacity. Consider that it may be necessary to shut down chillers slightly before the utility on-peak window (e.g., 12:45 AM in the previous example) and to turn on the chillers and associated equipment, one by one, a little after the on-peak window (e.g., 5:05 PM) to avoid unintended chiller operation during the on-peak demand period and stagger equipment startup.

In addition, typically, for backup purposes, in case a chiller is out of commission, one additional chiller equal to the size of the largest chillers is installed at the chiller plant (N+1 capacity).

Note that the minimum full storage thermal storage tank size (versus the previously calculated minimum chiller plant size) is the sum of the on-peak hours, which, in this case, is equal to 11,415 ton-hours as shown in Table 7-2. Note, however, that a larger chiller plant is required, 2668 tons, in this case, than in the previous example, to cover the hour before the chiller plant is turned off; since the thermal storage tank is already fully charged and there is nowhere to put the additional chiller output. Further, note that, with larger chillers than the minimum chiller size required, the chiller plant can be turned off for several hours during the early morning period. Also note that this minimum thermal storage tank capacity does not allow for some spare thermal storage capacity, as previously described, for example, at the end of the on-peak period while the chiller plant is brought on line. Nor does the minimum thermal storage tank capacity allow for complete chiller shutdown prior to the start of the utility on-peak period.

		Chilled water thermal storage system	
Time of day	Facility hourly CHW load (tons)	Hourly chiller output (tons)	Thermal storage charge inventory (ton-hours)
12:00 AM	1,090	2,668	6,441
1:00 AM	1,019	2,668	8,090
2:00 AM	959	2,668	9,799
3:00 AM	975	2,591	11,415
4:00 AM	969	0	10,446
5:00 AM	968	0	9,478
6:00 AM	1,274	0	8,204
7:00 AM	1,358	1,809	8,655
8:00 AM	1,457	2,668	9,866
9:00 AM	1,773	2,668	10,761
10:00 AM	2,176	2,668	11,253
11:00 AM	2,506	2,668	11,415
12:00 PM	2,668	2,668	11,415
1:00 PM	2,831	0	8,584
2:00 PM	2,900	0	5,684
3:00 PM	2,900	0	2,784
4:00 PM	2,784	0	0
5:00 PM	2,588	2,668	80
6:00 PM	2,348	2,668	400
7:00 PM	2,077	2,668	991
8:00 PM	1,945	2,668	1,714
9:00 PM	1,836	2,668	2,546
10:00 PM	1,711	2,668	3,503
11:00 PM	1,308	2,668	4,863
Total	**44,420**	**44,420**	
Thermal storage size (ton-hours)			**11,415**

TABLE 7-2   Peak-Day Full Thermal Storage Sizing Calculations—Minimum Tank Size

Given a required thermal storage capacity in ton-hours and the CHW system delta-T, the CHW thermal storage tank volume, using English units, can be calculated as follows:

$$V = \frac{X \cdot 12000 \, \dfrac{\text{Btu}}{\text{ton-hr}}}{c_p \cdot \Delta T \cdot SG \cdot 62.4 \, \dfrac{\text{lb}}{\text{ft}^3} \cdot eff}$$

where V = TES tank volume (ft³)

   X = amount of thermal capacity required (ton-hours)

   c_p = specific heat of water [1 Btu/pounds mass (lbm)-°F]

   ΔT = Temperature difference (°F)

   SG = specific gravity

   eff = Volumetric storage efficiency (0.9 typical)

   In the example from Table 7-1, assuming a 16°F delta-T, the thermal storage tank water volume = (12,641 ton-hours * 12,000 Btu/ton-hour) divided by (1 Btu/lbm-°F * 16°F * 1 * 62.4 lbm/ft³ * 0.9) = 168,817 cubic feet (ft³). Note that the typical tank rule-of-thumb volumetric efficiency is 90 percent, which accounts for the water volume above and below the diffusers, the external heat gain thermal losses, and the internal mixing or heat transfer losses associated with the thermocline. As discussed, actual thermal storage tank volumetric efficiency can be calculated or estimated and is a function of the water height, the thermocline thickness, the diffuser design (which can affect the thermocline thickness), the height of lower diffuser above the floor and the depth of upper diffuser below the water level, and the volume of any internal columns. Note also that some freeboard (air space above the water) is required and can be substantial in high seismic areas.

   With the thermal storage tank volume determined, the thermal storage tank dimensions can then be calculated. For example, using a cylindrical tank of 60 ft diameter, the minimum required water height in the proposed thermal storage tank would be approximately: tank water height = tank water volume divided by tank floor area = $V/(\pi*r^2)$, and in this case it equals 168,817 ft³ divided by [Pi * (30 ft)²] = 60 ft including a few inches for water thermal expansion (which equals the water level height times the maximum warm water density divided by the minimum cold water density).

   The actual tank shell height is equal to the water height (including thermal expansion) plus the required tank freeboard (e.g., to account for seismic wave slosh height). The total tank height equals the shell height and the height due to the tank roof slope to the center of the tank (equal to, e.g., ¾-inch per foot, times the tank radius). Of course, actual tank dimensions will depend on a number of factors including available area, soil conditions, internal column volume, amount of required insulation (a function of total surface area), construction feasibility, and tank construction cost economics, for example.

   It is important to note that full storage systems generally yield attractive economics for cases where one or more of the following characteristics apply:

1. Load profiles are very "peaky" (i.e., have short-duration peaks).

2. On-peak periods are relatively short.

3. Utility incentives (e.g., demand charges, on-peak versus off-peak energy charge differences, or demand-side management rebates) are high.

4. Large amounts of excess chiller plant capacity already exist.

Partial Storage (Load Leveling) System Sizing

Given the peak-day 24-hour cooling load profile, as shown in Table 7-3, the minimum chiller plant size needed to serve the facility cooling load is determined by dividing the

Time of day	Facility hourly CHW load (tons)	Chilled water thermal storage system	
		Hourly chiller output (tons)	Thermal storage charge inventory (ton-hours)
12:00 AM	1,090	1,851	1,455
1:00 AM	1,019	1,851	2,287
2:00 AM	959	1,851	3,179
3:00 AM	975	1,851	4,055
4:00 AM	969	1,851	4,937
5:00 AM	968	1,851	5,820
6:00 AM	1,274	1,851	6,397
7:00 AM	1,358	1,851	6,890
8:00 AM	1,457	1,851	7,284
9:00 AM	1,773	1,851	7,362
10:00 AM	2,176	1,851	7,037
11:00 AM	2,506	1,851	6,382
12:00 PM	2,668	1,851	5,565
1:00 PM	2,831	1,851	4,585
2:00 PM	2,900	1,851	3,536
3:00 PM	2,900	1,851	2,487
4:00 PM	2,784	1,851	1,554
5:00 PM	2,588	1,851	817
6:00 PM	2,348	1,851	320
7:00 PM	2,077	1,851	94
8:00 PM	1,945	1,851	0
9:00 PM	1,836	1,851	15
10:00 PM	1,711	1,851	155
11:00 PM	1,308	1,847	694
Total	**44,420**	**44,420**	
Thermal storage size (ton-hours)			**7,362**

TABLE 7-3   Peak-Day Partial Thermal Storage Sizing Calculations

facility total daily cooling load in ton-hours by 24 hours. Therefore, the minimum chiller size is the average of the peak-day cooling load and is equal to the area under the 24-hour load profile divided by 24 hours. Note, as with full storage systems, the total daily chiller production ton-hours must equal the total daily facility cooling load ton-hour demand.

In the example shown in Table 7-3, the facility daily ton-hours is again 44,420 ton-hours. Therefore, the minimum chiller plant capacity required to meet the facility's peak daily cooling load is approximately 1851 tons (44,420 ton-hours divided by 24 hours = 1851 tons). As noted in the previous full storage example, in practice, it is wise to provide some additional chiller capacity, depending upon how much of a safety factor is included in the estimated peak-day cooling load profile, as well as to provide N+1 chiller capacity in case one chiller is out of commission (or other redundant capacity, as may be desired by the facility owner).

Table 7-3 shows that the thermal storage tank charge for this example is 7362 ton-hours. A spreadsheet thermal storage sizing model is constructed in the same manner as described for the previous full storage example, except that, as noted, the chiller plant is operated at a continuous rate during the on-peak period.

Note that the on-peak chiller operation is more than a full storage case where no chillers are operated but less than a conventional chiller plant without thermal storage, which, in this example, would require 2900 tons of chiller capacity. Further, a chiller plant capacity of 1851 tons is the smallest and least expensive chiller plant that can meet the facility's peak daily cooling load of 44,420 ton-hours.

As shown in this sizing example, the CHW thermal storage tank is completely discharged of CHW by the end of the 8:00 PM hour (set the zero value at this hour). Starting at 9:00 PM, the chiller plant produces more CHW than is needed by the facility at that hour (1851 tons of chiller operation versus 1836 tons of demand). The net difference of the CHW production versus demand is stored in the thermal storage tank for use the following day. The chillers are operated at a constant output over the day, and, in this example, at the end of the 9:00 AM hour, the CHW thermal storage tank holds its maximum storage charge of 7362 ton-hours, after which the campus cooling load is greater than chiller plant output and the thermal storage tank begins to discharge to serve the campus CHW demand in excess of chiller plant output, until the cycle starts over again at 9 PM.

The thermal storage tank volume and dimensions can be calculated in the same manner as shown in full storage example, and the same considerations with regards to thermal storage tank volumetric efficiency also apply to partial storage systems.

Figures 7-3 and 7-4 show this example's partial thermal storage system operations graphically.

Hot Water Storage Sizing

Like all thermal storage systems, the size of the HW thermal storage system depends upon the following: (1) the peak load profile (HW in this case); and (2) the hours that the generating equipment will be producing HW. In some cases, for example, fuel fired HW boilers or cogeneration heat recovery units will operate continuously over a 24-hour period at maximum output in conjunction with HW thermal storage to meet the facility's HW demand. In other cases, HW generating equipment will only operate or produce heat during certain periods of the day, for example, solar thermal production only during daylight hours. In the case of cogeneration plants, HW thermal storage can allow for a cogeneration plant to operate more fully loaded than it might otherwise operate without dumping heat (which goes against the purpose of cogeneration). For example, during the day, heat can be stored for use later at night when the heat is needed.

To determine the required HW thermal storage capacity and corresponding required total heating plant capacity, like in the CHW thermal storage case, one can construct a

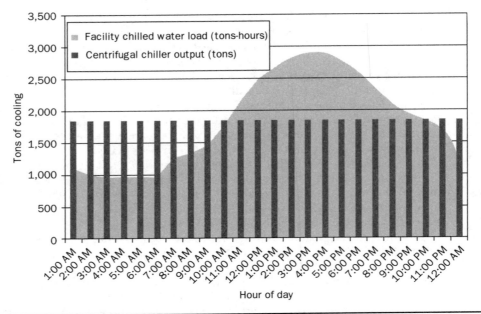

Figure 7-3 Partial thermal storage peak-day operation: facility cooling load vs. chiller output.

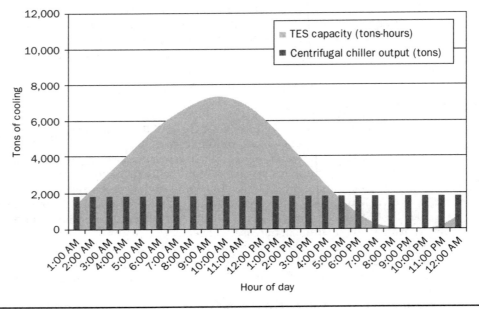

Figure 7-4 Partial thermal storage capacity and chiller output.

simple hour-by-hour spreadsheet model as shown in Table 7-4. The example shown in Table 7-4 provides the facility heating load profile; the thermal storage operating strategy (heat generating equipment off at night in this example); the resultant minimum required total HW generation equipment capacity, and the resultant minimum required HW thermal storage capacity. And, similar to a CHW thermal storage system, the required given information is the peak-day 24-hour heating load profile, which should

Time of day	Facility hourly HW load (MBtuh)	Hot water thermal storage system	
		Hourly boiler output (MBtuh)	Thermal storage charge inventory (MBtu)
12:00 AM	3,085	0	17,511
1:00 AM	3,440	0	14,071
2:00 AM	3,596	0	10,475
3:00 AM	2,131	0	8,344
4:00 AM	1,731	0	6,613
5:00 AM	1,676	0	4,937
6:00 AM	1,429	0	3,508
7:00 AM	3,508	0	0
8:00 AM	3,473	5,620	2,147
9:00 AM	3,445	5,620	4,321
10:00 AM	3,151	5,620	6,790
11:00 AM	2,763	5,620	9,647
12:00 PM	3,121	5,620	12,145
1:00 PM	3,018	5,620	14,747
2:00 PM	3,178	5,620	17,189
3:00 PM	3,217	5,620	19,592
4:00 PM	2,693	5,620	22,518
5:00 PM	2,985	5,620	25,153
6:00 PM	2,844	5,620	27,929
7:00 PM	3,051	5,620	30,497
8:00 PM	3,513	5,620	32,604
9:00 PM	3,807	0	28,797
10:00 PM	4,212	0	24,585
11:00 PM	3,989	0	20,596
Total	**73,056**	**73,056**	
Thermal storage size (MBtu)			**32,604**

TABLE 7-4 Hot Water Thermal Storage System Sizing with Equipment Off During a Portion of the Day

be on the day with the facility's maximum 24-hour total heat required [measured in thousands of Btu (MBtu) in this case] over a 24-hour period. The total daily MBtu is the sum of the day's 24 individual average hourly heating loads, and in the example shown in Table 7-4 it is equal to 73,056 MBtu (or 73 Mbtu).

Again, similar to the CHW example, the total 24-hour heating plant output must equal the facility's peak-day total daily MBtu demand. In this example, the heat generation equipment operates from 8:00 AM to 9:00 PM. Therefore, the required minimum heating plant capacity is calculated as approximately 5620 MBtu per hour (MBtuh) (73,056 MBtu divided by 13 hours, rounded up) and is entered into the spreadsheet sizing model as shown in Table 7-4. Note that minor heating output adjustments can be made to equalize the total daily MBtu. If actual heating plant capacity is or will be larger than the calculated minimum heating plant capacity required, one can use the capacity of the larger heating plant in the spreadsheet model.

The last column in Table 7-4 calculates the total thermal storage tank charge inventory for each hour; and the calculated minimum thermal storage capacity required is simply the maximum storage charge value calculated in the thermal storage tank charge column, which in this example is 32,604 MBtu. Similar to CHW thermal storage, the first step in calculating the maximum thermal storage tank charge is to identify the hour at which the thermal storage tank runs out of capacity and to enter a zero value in that cell. In this case, thermal storage capacity by design runs out the hour before the generation equipment restarts for the day.

Therefore, as shown in Table 7-4, the heat generation equipment starts to operate at 8:00 AM and the thermal storage runs out of capacity at the end of the 7:00 AM (i.e., 8:00 AM) hour after serving the load; and, therefore, the zero value is entered for thermal storage charge as shown. Again, similar to CHW thermal storage, the formula for thermal storage tank capacity for every other hour in the day is to take the previous hour's remaining thermal storage tank charge and add to it the difference between heating plant output and facility heating load for the current hour. When the difference between heating plant output and facility heating load is positive, the system is charging the thermal storage tank, and, conversely, when the difference between heating plant output and facility heating load is negative, the thermal storage tank is discharging to serve the facility's heating load demand.

Depending on how much of a safety factor is in the peak-day 24-hour heating load profile, it is probably prudent to add some additional capacity to the minimum thermal storage tank capacity and to the calculated minimum heating plant capacity.

Figure 7-5 shows this HW storage example graphically.

Table 7-5 shows the HW thermal storage case where the heating generation equipment operates at full capacity 24 hours per day. As expected, similar to CHW thermal storage systems, operating heating equipment over a 24-hour period results in the smallest required generation or production equipment and a smaller required HW thermal storage capacity than the case shown in Table 7-4.

Figure 7-6 shows this HW storage example graphically.

Ice Storage System Sizing

As in sensible storage, the first step for any ice storage system selection is to define the cooling load. In addition, as in sensible storage, latent (ice) storage may be either partial or full storage systems. Because of the greater energy needed to make ice than CHW,

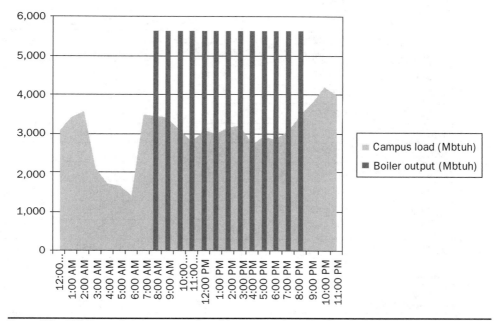

FIGURE 7-5   HW thermal storage operation with some equipment off part of the day.

Time of day	Facility hourly HW load (MBtuh)	Hot water thermal storage system		
		Hourly boiler output (MBtuh)	Thermal storage charge inventory (MBtu)	
12:00 AM	3,085	3,044	948	
1:00 AM	3,440	3,044	552	
2:00 AM	3,596	3,044	0	
3:00 AM	2,131	3,044	913	
4:00 AM	1,731	3,044	2,226	
5:00 AM	1,676	3,044	3,594	
6:00 AM	1,429	3,044	5,209	
7:00 AM	3,508	3,044	4,745	
8:00 AM	3,473	3,044	4,316	
9:00 AM	3,445	3,044	3,915	
10:00 AM	3,151	3,044	3,808	
11:00 AM	2,763	3,044	4,089	
12:00 PM	3,121	3,044	4,012	

TABLE 7-5   Hot Water Thermal Storage System Sizing with Continuous Equipment Operation

Time of day	Facility hourly HW load (MBtuh)	Hot water thermal storage system	
		Hourly boiler output (MBtuh)	Thermal storage charge inventory (MBtu)
1:00 PM	3,018	3,044	4,038
2:00 PM	3,178	3,044	3,904
3:00 PM	3,217	3,044	3,731
4:00 PM	2,693	3,044	4,082
5:00 PM	2,985	3,044	4,141
6:00 PM	2,844	3,044	4,341
7:00 PM	3,051	3,044	4,334
8:00 PM	3,513	3,044	3,865
9:00 PM	3,807	3,044	3,102
10:00 PM	4,212	3,044	1,934
11:00 PM	3,989	3,044	989
Total	73,056	73,056	
Thermal storage size (MBtu)			5,209

TABLE 7-5 Hot Water Thermal Storage System Sizing with Continuous Equipment Operation (*continued*)

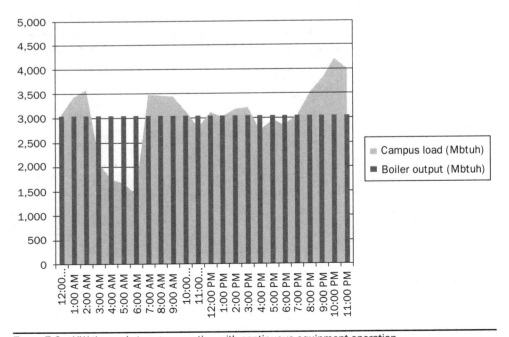

FIGURE 7-6 HW thermal storage operation with continuous equipment operation.

many partial ice storage systems produce ice only at night when other systems are off, and during the day they operate as standard chillers producing CHW for instantaneous use. When producing CHW, at a higher chiller operating temperature than when producing ice, the cooling equipment requires less power per unit of cooling produced.

Figure 7-7 is an example of an application with 11 hours of cooling load, from 7:00 AM to 6:00 PM, and a peak cooling load of 1000 tons. The total peak-day load is 9000 ton-hours, and this example assumes 10 hours of ice-building time. The rational for this operation is the utility rate with lower electrical costs off peak. For ice storage, the chiller is often assigned two operating supply temperatures, one supply temperature for the chiller's daytime capacity, when the chiller(s) is helping to meet the cooling load directly, and a second supply temperature for when the chiller(s) is producing ice at a colder operating temperature. When at ice-making coolant temperatures, most centrifugal chillers only produce approximately 70 percent (typically varies between 60 percent and 75 percent) of their nominal capacity that they produce when operating at typical building CHW system temperatures.

All of the cooling is produced by the chiller, and the basic approach is to equate the entire facility or building cooling load to the total contribution or production of the chiller, similar to the examples of sensible thermal storage systems sizing previously described. The chiller size required is complicated by the fact that far more capacity is available when operating as a chiller than when operating as an ice-making machine. The following calculates the minimum-sized chiller and storage capacity that meets the cooling load with the assumed conditions for this example:

$$TH_{total} = Q \cdot .7 \cdot t_{ice} + Q \cdot t_{cooling}$$

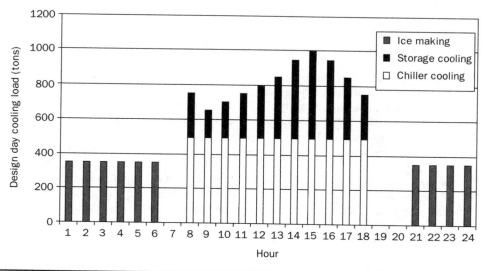

FIGURE 7-7 Minimum-sized chiller and storage contribution to cooling load.

And solving for the chiller capacity,

$$Q = \frac{TH_{total}}{0.7 \cdot t_{ice} + t_{cooling}}$$

where Q = minimum chiller (tons)

TH_{total} = design day total cooling load (ton-hours)

t_{ice} = time chillers are making ice (hours)

$t_{cooling}$ = time chillers are helping to meet load directly (hours)

0.7 = Ratio of ice-making capacity to conventional chiller capacity (system dependent)

Once the chiller capacity is determined, the amount of storage, in ton-hours, can be calculated.

$$TH_{storage} = Q \cdot .7 \cdot t_{ice}$$

Substituting the known values for this example results in a minimum chiller size of 500 tons that provides 350 tons of ice-making capacity and requires 3500 ton-hours of ice storage capacity, providing a potential electrical demand reduction corresponding to 500 tons of daytime chiller operation as shown in Fig. 7-7.

Although this simple analysis defines the minimum requirements, other factors, such as chiller plant redundancy, night load, and space limitations, often influence the design and equipment selection. However, the calculation procedures follow a similar path. The total cooling requirement is equated to the contribution of the chiller(s) with the proper assignment of operating hours and associated capacity multiplier (e.g., 0.7 for ice making). Keep in mind that, as the selected chiller capacity exceeds the previously calculated minimum, the larger chiller can reduce the need for storage because it can contribute more cooling production during the day, or the additional chiller capacity can produce even more storage. Of course, full storage represents the highest chiller capacity design limit. In this case, 9000 ton-hours of storage would provide the entire daytime cooling load, with a chiller(s) of 900 tons ice-making capacity (approximately 1300 tons nominal, but not used in this example) running only during the 10 off-peak nighttime hours. Dividing the chiller capacity among two or more machines usually provides all the desired redundancy for full storage, since the system can often simply revert to partial storage operation with the loss of a single chiller.

For partial storage, redundancy requirements are usually addressed differently. A common conventional approach, without thermal storage, is to add 20 percent to the design day peak load (e.g., 1000 tons * 1.2 = 1200 tons) and divide the total into three equal capacity machines (e.g., 3 × 400 tons). Note that by substituting thermal storage for the third chiller, similar levels of redundancy can be achieved in addition to the benefit of electric demand reduction. As previously noted, the increased chiller capacity, above the theoretical minimum, also increases the amount of storage that could be applied.

Some options for this example that also address redundancy, while employing 2 × 400 ton chillers, include the following:

- Operate two nominal 400-ton chillers capable of producing up to 5600 ton-hours of stored ice cooling in 10 hours, which would allow, on the peak cooling day, the daytime chiller contribution to be reduced to one machine at a part load of approximately 310 tons, reducing on-peak cooling demand by 690 tons (the total daily ton-hours on the peak cooling day is 9000 ton-hours, and since 5600 ton-hours is provided from ice storage 3400 ton-hours must be produced over the 11 operating hours, or 309.1 tons; therefore, since the conventional peak load is 1000 tons, but only 310 tons of chiller is required, the electric demand associated with cooling production is reduced by approximately 690 tons in this example).

- Operate a single 400-ton chiller at full capacity during the day, reducing the required thermal storage capacity to 4600 ton-hours (9000 − 400 ∗ 11), while still eliminating 600 tons of on-peak chiller-related electrical demand.

- Select the amount of storage that a single chiller could produce in the 13 no-load hours or 3640 ton-hours (400 ∗ 0.7 ∗ 13). Note that the single chiller system capacity would be 8040 ton hours (400∗11 + 3640) or almost 90 percent of the design day ton-hour cooling requirement—almost equal to the redundancy provided by the 3 × 400-ton chiller, conventional system.

The last option would require 490 tons of daytime chiller contribution, almost equal to the minimum-sized system first discussed, which is very close to the original partial storage approach with a minimally sized chiller as shown in Fig. 7-8.

Other than the minimally sized partial storage system first described, and a full storage design, there are also a number of acceptable alternatives or variations.

The preceding analysis determines the chiller sizes and storage capacity required for a given peak-day cooling load profile. The ice storage equipment capable of delivering that capacity must then be selected. External melt storage capacities are relatively straightforward, as the quantity of available ice storage is, to a large extent, only a function of total ice inventory. As discussed in Chap. 5, the performance of

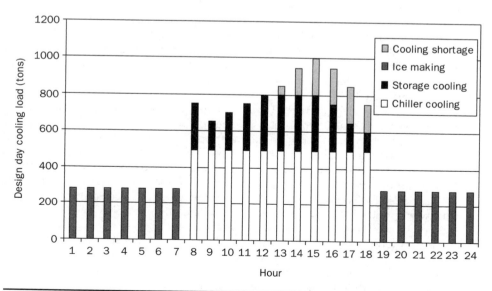

Figure 7-8   Minimum-sized chiller and storage contribution to cooling load.

internal melt storage products is more dependent on coolant temperatures, flow rates, and storage inventory (amount of ice remaining). System design will also influence the ice storage performance. Supply and return temperatures and the arrangement of components also influence the ice system operating capacity.

A common primary or secondary configuration places the chiller and ice storage in series, with the chiller upstream or downstream of the storage equipment (versus sensible heat storage where the chiller and storage are virtually always configured in parallel). With an internal melt ice storage system, the coolant temperature leaving the ice storage equipment varies throughout the day as cooling loads change and ice inventory diminishes, and the manufacturer must ensure that the ice thermal storage can provide the discharge rates and the proper coolant temperatures throughout the design day.

The operation depicted in Table 7-6 assumes a chiller upstream of storage with design supply and return temperatures of 42°F and 58°F respectively. The thermal storage must always be able to provide 42°F coolant temperatures to the facility or building in order to properly serve the cooling loads. Too cold supply temperatures may cause problems of excessive condensation and dehumidification of supply air. In addition, lower supply temperature will usually result in lower return temperatures and that would unload the chiller. Consequently, in this example, a bypass valve is needed to divert some supply coolant around the ice module to provide the desired 42°F leaving coolant, which means the chiller meets the load to the extent possible, and, when the load exceeds the chiller capacity, the additional load is supplied from the ice storage.

If there is no problem with the supply coolant being colder than design, then the bypass can be eliminated and the chiller leaving coolant temperature (LCT) set for the temperature desired to make ice. The chiller operates to cool the coolant to the temperature required to make ice, or, if the return temperature is higher than design, to the temperature allowed by the chiller capacity. When the temperature leaving the chiller is below the fusion point (32°F for ice), all of the building load is met by the chiller and some ice is produced. When the temperature leaving the chiller is above 32°, any load not met by the chiller is provided by melting some ice.

The chiller contribution is limited to 500 tons during the on-peak daytime cooling period as illustrated in Fig. 7-7. For a typical primary or secondary design, the following table summarizes the hourly conditions on a design day after the proper amount of storage has been determined.

In this example, the storage LCT is a maximum (41.1°F) during the last cooling hour of the day, that is, during the worst-case hour. Since these types of ice storage systems use a heat exchanger to separate the ice from the circulating coolant, there is almost always excess ice storage capacity available, although it may not be delivered at the design temperatures. Ice storage systems should always be designed to provide, at least, the worst-case design day performance requirements. Note that, during the day, the chiller LCT is considerably higher than it would be in a conventional system, thus illustrating one of the factors that helps balance the efficiency penalties of reduced temperatures needed to produce the stored ice.

Even though the cooling load provided by ice storage is half of the peak value, the reduction in ice inventory dominates the performance of the ice thermal storage system. In some cases, it may be determined that an earlier hour, with higher storage cooling loads but greater ice inventory, is the critical design point.

Hour	Cooling load (tons)	Chiller (tons)	Storage (tons)	Storage inventory (ton-hours)	Return temp (F)	Chiller LCT (F)	Storage LCT (F)
1	0	331	331	2,316	31.0	25.0	31.0
2	0	331	331	2,647	30.6	24.6	30.6
3	0	331	331	2,978	30.0	24.0	30.0
4	0	331	331	3,309	29.1	23.1	29.1
5	0	331	331	3,640	28.0	22.0	28.0
6	0	0	0	3,640	—	—	—
7	0	0	0	3,640	—	—	—
8	750	500	−250	3,390	54.0	46.0	32.5
9	650	500	−150	3,240	52.4	44.6	32.6
10	700	500	−200	3,040	53.2	45.0	32.7
11	750	500	−250	2,790	54.0	46.0	32.9
12	800	500	−300	2,490	54.8	46.8	33.5
13	850	500	−350	2,140	55.6	47.6	34.6
14	950	500	−450	1,690	57.2	49.2	36.7
15	1,000	500	−500	1,190	58.0	50.0	38.7
16	950	500	−450	740	57.2	49.2	39.1
17	850	500	−350	390	55.6	47.6	38.7
18	750	500	−250	140	54.0	46.0	41.1
19	0	331	331	331	31.8	25.8	31.8
20	0	331	331	662	31.7	25.7	31.7
21	0	331	331	993	31.7	25.7	31.7
22	0	331	331	1,324	31.6	25.6	31.6
23	0	331	331	1,655	31.5	25.5	31.5
24	0	331	331	1,985	31.3	25.3	31.3

TABLE 7-6   Design Day Ice Storage Operating Strategy Example

Summary

As shown, thermal storage sizing is a function of the following:

- Type of thermal storage system
- Thermal load profile
- Thermal strategy employed (partial storage versus full storage)
- Number of hours thermal production equipment is operated

- Number of hours thermal production equipment is off
- Size of thermal production equipment
- Delta-T (for sensible thermal storage systems)
- Tank geometry

Ice storage systems are sized somewhat differently from CHW thermal storage systems, in part, because of the inherently different chiller operating capacity, depending upon whether the chiller is producing ice or relatively warmer CHW.

CHAPTER 8

Conducting a Thermal Storage Feasibility Study

Introduction

Conducting a thermal storage feasibility study is a necessary step to determine the technical feasibility and potential benefits of a thermal storage installation. A good thermal storage feasibility study should consider (and, where appropriate, examine) a number of different possible types of thermal storage systems (e.g., water and ice for cooling applications) as well as different thermal storage operating strategies (e.g., partial storage and full storage). These different types of thermal storage and different types of thermal storage operating strategies should be contrasted against a business-as-usual (BAU) or conventional case(s) without thermal storage. The purpose of a feasibility study is to determine the possible economic or operational benefit of a thermal storage installation. Economic benefit includes lower operational energy cost and potential construction cost savings. Operational benefit may include the ability to serve a load for a period of time without the operation of a conventional cooling or heating system.

The technical feasibility and economic performance/viability of various options/alternatives need to be considered. Usually the recommended option is the option with the lowest life cycle cost (LCC) or greatest LCC savings. There are, however, many exceptions including where higher reliability is critical, where the ability to meet future expansion is needed, or where the ability to meet expansion needs within a certain footprint is required. Every study/project is, of course, somewhat unique.

A thermal storage feasibility study provides a process that is the basis for solid engineering and financial recommendations. That is true if a project is a simple expansion of a small building but is especially true for a project that may cost millions of dollars to construct.

First, the feasibility study author should understand the facility owner's goals and reasons for considering thermal storage, as the reason may affect to a large extent the feasibility study itself. Some examples of different reasons to consider thermal storage include thermal service to a new facility or replacing an old and/or failing existing thermal system (e.g., steam or hot water). Alternatively, the reason may be to allow for future facility expansion or even to provide for emergency thermal supply. Often the reason is one of pure economics; that is, can the owner save money over time by the installation of a thermal storage system, and, if so, do the estimated savings justify the cost of the thermal storage installation?

The feasibility study author also should review any previous reports and/or studies as well as records that may be applicable to the thermal storage feasibility study. As outlined below, to gain a good understanding of the existing thermal production and use systems, a thorough review of any available as-built drawings followed by on-site field investigation to review/verify those drawings are required. In addition, it is always helpful to meet and talk with facility personnel in order to gain a better understanding of facility and project goals, to gain an insight into any operating challenges and/or issues, as well as to gain a better understanding into the condition of existing systems and equipment. Moreover, a key part of any project is the buy-in of the facility operating personnel.

For existing facilities, an important part of a feasibility study is the field data collection. Some of the factors that should be gathered on site are as follows:

- Peak load data for existing equipment: flow, delta-T, and pressure differential
- Coil: delta-T, load, and pressure drops (water and air side)
- Elevation of the highest coils (this affects some potential solutions)
- How building or process loads are controlled and the ability of control valves to work against potential pressure differentials
- Pressure difference in piping systems at peak loads

Many of the above factors have a significant bearing on the potential solutions ultimately recommended. As an example, it may be more economical to replace coils in a poor performing system (i.e., coils with a low delta-T) than to design a larger thermal storage tank. Moreover, it may be more economical to take steps to increase system delta-T than to oversize pumps or to install larger pipes. With mechanical/thermal systems, there is nearly always more than one way to solve a problem, and the key challenge is to really understand the problems and the potential solutions.

The next key step in the thermal storage feasibility process is to develop/estimate the facility peak existing and/or future thermal load profile for use in sizing the thermal storage system. In addition to a detailed understanding of the existing system (if any), the study must include any planned/future expansion. That means estimating thermal loads for buildings or processes that do not exist. Often little more is known than a potential building size or process load and when the added/increased load will come on line. Nevertheless, as good an understanding as possible of the existing and future loads is another key element in a thermal storage feasibility study.

Given the facility peak thermal load profile, the engineer(s) develops technically feasible options to meet the thermal load (cooling load or heating load). Note, the number of options to study is normally agreed upon as part of the engineering study proposal.

Once the possible options to consider have been developed, the next step is to investigate the technical feasibility of those options. It may be likely that some options will drop out from the study as they are found to be not technically feasible. For example, there is no room for an aboveground tank and only shallow underground tanks can be used, or underground tanks are not feasible because of the hard rock or the high water table and only aboveground tanks can be used. As part of the investigation into the technical feasibility of the various options, the engineer develops a conceptual mechanical system plan for each option/alternative. With a concept plan developed for each concept, budget construction costs as well as initial annual maintenance costs can be estimated.

In addition, with a concept plan developed, the next step in the feasibility study process is to estimate energy use and energy costs for each option/alternative, typically, using energy software or using the spreadsheet method. Once the construction cost, annual energy costs, and annual maintenance costs have been estimated, the next step in the thermal storage feasibility process is to perform an economic analysis using engineering economics. The last step in the feasibility study is to make project recommendations based on the technical and economic feasibility.

The typical thermal storage feasibility study scope of work includes the following:

- Meet with the facility representatives/owner(s) to review: the reasons for conducting the thermal storage feasibility study; project goals; schedules; agreed scope of work; and communication protocols.
- Obtain and review available architectural master plan (get an understanding of any future expansion plans).
- Obtain and review facility thermal production and use equipment as-built drawings.
- Obtain and review utility consumption data.
- Obtain and review existing and available utility tariffs.
- Identify and consider any available utility rebates.
- Conduct on-site facility field investigation.
- Review and analyze existing thermal plant equipment.
- Estimate existing facilities loads, as applicable.
- Estimate future facility loads, as applicable.
- Develop a concept-level BAU base case alternative.
- Develop concept-level thermal storage system options and requirements.
- Use existing energy bills to establish the BAU annual energy use and cost, and/or estimate the BAU annual energy use and costs.
- Estimate thermal storage annual energy use and cost for each technically feasible option/alternative.
- Prepare concept-level budget construction cost estimates for each option/alternative and estimate total project costs.
- Conduct basic sensitivity analysis for higher delta-T on sensible chilled/hot water thermal storage systems.
- Calculate economics/LCC for each option/alternative.
- Prepare a draft and final report of all findings.
- Make an executive summary presentation to key stakeholders.

The following is a sample table of contents for a thermal storage feasibility study:

1. Executive Summary
2. Background
3. About Thermal Storage
4. Existing Systems
   a. Delta-T
   b. Elevations
   c. Controls
5. Existing and Future Thermal Loads
6. Thermal Storage Options
   a. Identify and describe various options
   b. Advantages and disadvantages of various options
   c. Any options considered but ruled out and the reason why
   d. Options selected for additional evaluation
7. Estimated Energy Use and Costs
   a. Energy rates available
   b. Projected energy costs with various options
   c. Consideration of future trends in energy cost
8. Estimated Budget Construction Costs
9. Economic Analysis
   a. Estimated maintenance costs
   b. Time value of money
   c. Escalation and depreciation factors
   d. Additional (secondary) benefits and value of thermal storage, if any
10. Recommendations

The executive summary, of course, summarizes the key points of the entire report so that someone can quickly understand the issues, results, and recommendations. The background chapter provides the reasons for considering the thermal storage systems, a brief overview of any expansion plans, a description of any relevant key issues affecting the facility thermal production and use systems and/or equipment, as well as general information about the facility itself. The chapter on existing systems should include key equipment schedules with all relevant capacity data as well as a review of equipment condition (excellent, good, fair, poor, failing, inoperative), and can act as an information resource for engineers and facilities personnel.

The chapter in the study on existing and future loads is, of course, a key chapter since, as previously discussed, the facility peak thermal load profile is the key input into thermal storage system sizing and should be reviewed carefully by all parties. The following pages in this chapter describe developing thermal storage system alternatives, estimating energy use and cost, estimating budget construction and project costs, and conducting an economic analysis.

Developing Thermal Storage System Alternatives

As discussed, in order to develop thermal storage alternatives, the peak thermal load profile must be known, and one must have a good understanding of the facility: needs and goals, expansion plans, existing thermal production and use systems, and available space. With the peak thermal load profile known, the required thermal storage and production equipment capacity can be determined as shown and described in Chap. 7. Also, as noted, the number of options/alternatives to analyze should be agreed to as part of a proposal process and is usually a minimum of three options/alternatives. The first alternative to develop is a conventional or BAU case.

The BAU case is the case represented by the normal or typical or least cost way/method of producing and using thermal loads, and sometimes represents the "do nothing" case (i.e., just keep existing systems). An example of a BAU case might be using existing thermal equipment and/or installing new thermal generating equipment (e.g., boilers and/or chillers) to meet the facility thermal loads. If existing equipment is planned to be used in the BAU case, consideration must be given to the age and condition of the equipment. If the equipment being used in the BAU case has a remaining useful service life less than that of the alternatives to which it is being compared, then replacement of the existing equipment must be factored into the LCC analysis of the BAU case. Similarly, if the existing equipment will need major overhauls (outside of that to be considered in the annual maintenance costs), then those costs need to be factored into the LCC of the BAU case as well. The BAU is also often used to calibrate models and assumptions before use with the other thermal storage options/alternatives.

Brainstorming is a good method for developing thermal storage options/alternatives for consideration. Brainstorming involves recording all ideas without criticizing or editing, even those ideas that might not be technically feasible, with the idea to maximize the number of ideas recorded. In fact, sometimes, including "crazy" ideas during the brainstorming process allows other good ideas to come through for inclusion. After the brainstorming process is complete, the best ideas can be selected from the brainstorming list for further investigation and analysis. Some thermal storage options to consider may include the following:

- Chilled water thermal storage
  - Vertically stratified
  - Multiple tank
  - Water with special additives (low temperature fluids)
  - Dual-use of thermal storage as a fire protection reservoir

- Ice thermal storage
  - Internal melt ice-on-coil thermal storage system
  - External melt ice-on-coil thermal storage system
  - Encapsulated ice thermal storage system
  - Ice harvester thermal storage system
  - Ice slurry thermal storage system

- Hot water thermal storage
- Gravel or earth/ground thermal storage
- Building mass thermal storage

- Partial storage system
- Full storage system
- Aboveground thermal storage tank(s)
- Belowground thermal storage tank(s)

Additional options/alternatives to consider often include different thermal storage tank sizes/shapes, as well as different potential thermal storage locations around the facility, and different thermal storage operating (charge and discharge) temperatures. Given the possible options/alternatives to consider, the technical feasibility of those options are investigated, and some options may drop out of consideration if determined not to be technically feasible, or if known to be excessively expensive and/or beyond the owner's project budget, or for other reasons.

As part of the investigation into the technical feasibility of the various options, the engineer develops a conceptual mechanical system(s) for each option, which typically includes selecting and sizing the number and type of equipment including thermal storage tank size and generation equipment size; developing a simple flow diagram to show concepts and number and size of equipment; and preparing a concept-level equipment layout (plan view). The engineering effort includes making contacts and having discussions with equipment vendors.

Selecting the right number of equipment to use is somewhat of an art and often depends on a number of factors including the requirement for equipment backup, the size of the equipment, the new equipment costs, the installation costs, energy performance, and the space available. If backup of the largest installed production equipment is required (i.e., $N + 1$ redundancy), then using multiple units is often more cost-effective than having to provide a backup for one large machine sized for the total thermal load (unless equipment size is relatively small). Multiple units also can allow the facility part load to be served more effectively and efficiently. However, installing multiple units is typically always more expensive than installing a single unit of the same combined capacity; and at some point the number of units becomes excessive.

With fairly high delta-T systems, there is a significant savings with series chiller machinery because only the final equipment must provide the final (lowest) discharge temperature. As an example, consider two chillers in series providing chilled water from 65 to 40°F. The lead chiller may only need to cool from 65 to 52°F, while the second chiller cools from 52 to 40°F. Because the lead chiller does not have to provide as high a "lift" to the condensing temperature, it uses less energy than the final chiller. A single chiller would need to provide the full lift and uses about the same energy consumption per unit cooling capacity generated as the final chiller. If all other factors are held constant with 95°F condensing temperature, one can expect the lead chiller to use about 20 percent less energy per unit capacity than the second/final or the single chiller.

As discussed in previous chapters, the size of a sensible water thermal storage tank (chilled water or hot water) is inversely proportional to the delta-T: the larger the delta-T, the smaller is the required thermal storage tank size. Additionally, lower hydronic flows are needed with higher delta-Ts to meet the same thermal load than with lower delta-Ts; therefore, smaller pumps and smaller distribution piping are required, with subsequent lower required capital cost and lower pumping energy. Therefore, maximizing the facility delta-T is an important factor in thermal storage feasibility, and, if required, the feasibility study author should, at least, recommend considering increasing the facility delta-T and including higher delta-T analysis as part of the feasibility study. In addition to capital cost issues, as higher delta-Ts allow for smaller thermal storage tanks for a given thermal storage capacity, higher delta-Ts can make it easier to site thermal storage systems.

Estimating Energy Use and Cost

The largest portion of the annual costs in LCC analysis is typically the energy costs. With cooling thermal storage systems, a substantial amount of estimated energy cost savings results from electric demand charge savings and from shifting the time when energy is used. Unless there is no change in facility use and good data on energy use is available, the energy usage must be estimated by computer modeling. Typically, this is done using energy software. Simple analysis may be done by using an existing load profile and adjusting it for the expected expanded load. The estimated energy use and cost must be estimated for each technically feasible option included in the thermal storage feasibility study. Energy costs are calculated based on the magnitude and timing of energy consumption, given utility company pricing tariff structures. The pricing structures for electricity and gas required in operating building systems and thermal plants are, as discussed in Chap. 2, often complex and vary with the season and the time of day. Electricity is often priced on an energy use and demand basis with different rates applying during different times of the year (or season) and different times of the day. And as discussed, electricity rate schedules commonly include a demand charge that is based on the maximum power use during a predefined period of time. Natural gas is usually priced on an energy basis ($/therm or $/MMBtu) with different rates applying during different times of the year (e.g., winter vs. summer).

Several commercially available software programs are able to estimate energy use load profiles. However, the software programs are sometimes far less able to model performance of specific thermal storage solutions, which are often developed to meet specific needs or utility rate structures. The best solution is sometimes to use the computer model to determine hourly loads and then to use spreadsheet solutions to model how the various thermal storage systems will meet that load. Advantages to software models are ease of use, repeatability, and presumed quality assurance of the model. Disadvantages of some software models may include limitations in modeling unique thermal storage applications, the model may have flaws that are less than initially apparent to the user, or require "tricking" the simulation model by adjusting the input to model an aspect of the thermal storage scenario that the model does not appear to support.

Spreadsheet models can be built that allow for very detailed analysis of unique or "out of the ordinary" thermal storage applications, or allow the modeling to report a unique aspect of the results. Advantages of spreadsheet modeling include the ability to build as detailed and unique a model as the user desires and increased ability to follow the logic behind the calculations. Disadvantages of spreadsheet models include more time to initially create and check the model, a higher chance of errors if close attention is not paid to detail, a lack of annual hour-by-hour calculation, and making even small changes to the model once it is substantially complete, which can prove challenging.

Note that as the thermal storage system is sized for the peak day thermal load, during other periods of lighter loads the operation of the thermal storage system can and should be adjusted to minimize utility cost generally by maximizing off-peak operation. So, for example, a thermal storage system designed as a partial thermal storage system on the peak day can be operated as a full storage system on light load days. Any energy model must take this fact into account.

Estimating the Budget Construction Costs

The budget construction cost/total project cost (i.e., the first cost/capital cost) is often a key factor that determines whether a project will proceed or not. If the construction cost is too expensive relative to the estimated energy savings, the project will likely not proceed. The project costs include engineering and project administration costs and contingencies, for example, as well as construction costs. The project costs/first costs are a key input into the LCC analysis. Depending on the depth of the LCC analysis being prepared, budget construction cost estimates can range anywhere from a cost per ton-hour basis, for example, to a detailed item-by-item cost estimate basis. Resources for preparing the budget construction cost estimate include equipment manufacturer/ vendor quotes, cost estimating publications, contractor estimates, and professional cost estimators. It is advisable to always obtain multiple vendor quotes for all major equipment.

Part of preparing an accurate budget construction cost estimate is having a good understanding of all the pieces required for complete construction. Other important components of the budget construction cost estimate to consider in addition to the equipment, material, and labor costs are as follows:

- *Subcontractor markup.* Oftentimes, cost estimating publications provide the raw material and labor costs. The actual "burdened" costs of labor can be much higher when all of the payroll taxes, benefits, and so on for the workers are considered. Markups of 10 percent for materials and up to 50 percent for labor are common.

- *Location factors.* Most cost estimating publications provide location factors to adjust for the higher or lower costs of materials and labor in the area where the project is located. In the United States these may be in the range of ±10 percent and are added to the subtotal of the costs estimated from the cost estimating publication (with subcontractor markups).

- *Open shop.* Requirements for union versus nonunion construction labor can also have a significant cost impact.

- *Taxes.* Sales tax may be applicable to all purchased material depending on the locale and is applied to the subtotal including the above markups and location factors.

- *General requirements.* General requirements cover the contractor's cost of reproduction, office equipment, constructions trailers, mobilization and demobilization, project management, and so on. Typical values are often around 5 percent and are either estimated individually or the 5 percent factor is applied to the subtotal including the sales tax.

- *Contingency.* A contingency amount should be added to cover unexpected costs. Contingencies may range from 5 percent for a very detailed cost estimate based on final engineering drawings to 20 percent for a rough "order of magnitude" type cost estimate based on concept-level ideas. The contingency is applied to the subtotal including the general requirements.

- *Insurance and bonds.* The cost of the contractor's insurance and bonds is typically around 3 percent and is applied to the subtotal including the contingency.
- *Contractor's overhead and profit.* This typically ranges from 10 to 15 percent and is applied to the subtotal including the insurance and bonds; however, the actual percentage is dependent on the overall health/status of the economy as well as on local market conditions.
- *Owner's project costs.* In addition to the budget construction costs, additional project costs should be added to the budget construction cost, such as the cost of engineering design, construction testing, construction inspection fees, and the owner's construction administration. Project costs are typically about 20 percent of the construction costs.

Estimating Annual Maintenance Costs

Another important factor in an LCC analysis is the annual maintenance costs, which must be calculated for each thermal storage alternative as well as the BAU case. Typical maintenance costs can include

- Preventative/periodic maintenance of equipment
- Costs of consumables
- Repair of equipment
- Rebuilding/overhaul equipment during the life of the analysis
- Cost of operators and maintenance personnel
- Cost of facility administrative staff

Costs for operations and maintenance staff, as well as administrative staff, can vary largely from region to region and from facility to facility. Some facilities may already have these staff positions covered, whereas some facilities will have to start from the beginning in staffing their operation. Most thermal storage facilities may be designed to operate with minimal operator supervision/intervention.

Most major equipment manufacturers can offer historical maintenance and repair costs. Oftentimes, maintenance contracts are available that provide all necessary maintenance and repairs on either a flat annual rate or based on hours operated. Additionally, publications are available that offer typical maintenance and repair costs based on surveys of equipment already installed and operating at various facilities.

Chilled water and hot water thermal storage tank maintenance is typically minimal, and concrete tanks typically require the least maintenance. Aboveground steel thermal storage tanks should be visually inspected annually (note chilled water treatment testing should be conducted on a weekly, if not daily, basis, at least during initial operation until satisfactory results allow for less frequent testing), with special attention paid to the vapor space above the water level as this is where most corrosion is likely to occur. It is possible that the thermal storage tank itself will need little (painting touchups) or no maintenance for several decades.

Conducting Economic Analysis

Economic analysis is the process by which the financial attractiveness/viability of each alternative is determined for comparison and contrast. The criteria for defining a project as economically viable will vary from project to project, for example, a minimum rate-of-return or a maximum number of years to pay back the investment. Economic analysis methods vary from a simple payback analysis to the more complicated and detailed rate-of-return or net present value (NPV) LCC analysis discussed in detail further in this chapter.

Simple payback analysis is an economic analysis method that looks at the time required to recoup the first costs based on the initial annual savings realized from the installation of the project. Simple payback analysis does not account for the time value of money, escalation, and so on, as does LCC analysis. Use of simple payback analysis is typically limited to the initial feasibility stages of the project. As more detailed study of the project is performed, LCC is usually required. The formula for simple payback is a follows:

$$Simple\ payback\ (years)\ =\ \frac{project\ first\ cost}{annual\ savings}$$

where *project first cost* = the total installed cost of the project including engineering and administration costs

annual savings = the sum of the savings in energy, operations, and maintenance costs compared to the BAU.

The project first costs in the formula preceding can also be the difference in cost between alternatives divided by the difference in savings, which can help answer the question of whether the added investment is justified or not.

LCC analysis is a process by which different first costs, different ongoing escalating annual costs, and different salvage values can be compared and contrasted on an equivalent basis. LCC analysis allows one to compare not just the first costs, and not just the proposed energy costs, for example, but all costs over the life of a project/system. LCC takes into consideration the time value of money and the escalation of various costs; it is also known as engineering economics, which allows one to answer the question: is it worth it to invest money now for the return in the future? Therefore, by developing the LCC of each alternative under consideration (including the conventional or BAU case), one can determine the comparative advantage of the proposed system in terms of receiving the best economic return on the money invested [also called rate of return on investment (ROI) or internal rate of-return (IRR)]. The following paragraphs define the economic terms for components of an LCC analysis and describe the process of performing an LCC analysis, including some examples.

LCC Process

LCC analysis takes the many factors that affect the economics of potential solutions and develops a single number (NPV) that can be compared and contrasted. The factors include the cost or value of money to the owner, inflation, depreciation, operating costs, as well as the project costs and energy costs. For each alternative, the feasibility study author should determine the engineering economics as follows:

1. Estimate the first costs (budget construction and project costs); the annual costs (energy, maintenance, financing, taxes, etc.) escalated for inflation; and the salvage value (if any) of systems and equipment at the end of the economic analysis period (e.g., 20 years).
2. Construct a cash flow model.
3. Calculate the resultant NPV or equivalent uniform annual cost (discussed later in the chapter).

The alternative with the lowest NPV of costs (or highest NPV of savings) will be the best alternative in economic terms. If long-term cost savings, as represented by a positive NPV of annual savings, are obtained with a chosen interest rate and realistic selection of economic factors, including escalation, that alternative is considered to be economically viable relative to the BAU case.

Capital Costs versus Annual Costs

Capital costs are those associated with constructing the thermal storage system including the thermal storage tank/system, the purchase and installation of all necessary equipment, controls, instrumentation, piping, and appurtenances needed for operation, the construction of a plant building. Typically, capital costs are the first costs of an alternative; however, capital costs can occur further into the project (i.e., not at year 0), if replacing existing equipment that is at the end of its service life or replacing equipment with a service life less than that of the entire project is required.

Annual costs are those that go toward operation of the thermal plant and are considered expenses. Examples of annual costs are electricity purchases and fuel purchases from the utility, labor costs for operators, cost of consumables, administration costs, and periodic maintenance and repair costs. Some maintenance items, repairs, or replacements happen at a frequency of more than 1 year. In this case, the annual cost calculation may "annualize" those costs. For example, if the project life is 20 years, but every 8 years some equipment needs to be rebuilt, the equivalent uniform annual cost of the rebuild cost (on an escalated basis) is budgeted and set aside every year.

Cash flow diagram

Cash flow is often a very necessary issue to an owner and takes into account not only how much money will be needed but when the money will be needed. In some cases, a solution with a higher LCC may be preferred, if it defers when cash will be needed, because of when future equipment will need to be on line. Cash flow (also called net cash flow) shows first costs (or investments) and periodic individual and total costs (or payments) for a given time period. Table 8-1 shows anticipated net energy savings following construction and beneficial use.

A cash flow diagram represents a cash flow table in a diagrammatic manner using a horizontal line to represent the time periods, upward facing arrows for receipts (positive cash flows), and downward facing arrows for payments/costs (negative cash flows).

Time Value of Money

The premise of "time value of money" is that, due to escalation (inflation) and potential interest earnings, the same money received in the future is worth less than that same

Time period	First costs ($)	Energy savings ($)
Year 0	2,300,000	0
Year 1	0	500,000
Year 2	0	525,000
Year 3	0	550,000
Year 4	0	575,000
Year 5	0	600,000

TABLE 8-1   Sample Cash Flow Chart

money received now. That is, $1,000 today is worth more than $1,000 in 10 years. Therefore, an investment that will save $1,000 in the first year is worth more than the same investment resulting in saving $1,000 10 years later. If an 8-percent return on investment (the discount factor) is available for 10 years, then $1,000 in 10 years is worth $463 in today's dollars (or has a present value of $463). Conversely, if that same $463 was invested at 8 percent for 10 years the final value will be $1,000.

Discount Rate
The discount factor is one of the most important numbers used in LCC analysis. The number used as the discount factor equates future values with present values. That is, the discount factor is the number used to determine the equivalent present dollar value given some future dollar value. In general, the discount factor should equal the long-term cost (or value) of money or the alternative investment rate. Higher discount factors will "discount" (devalue) future values even more than lower discount factors. That is a greater discount factor will reduce the importance of future costs (or savings) in the economic analysis.

Interest Rate
The interest rate is the rate paid by a borrower for the use of someone else's money, or, alternatively, the return a lender receives for deferring the use of their money to lend it to a borrower. If the capital to construct a project is borrowed, the interest rate will be paid to the lender. If the money is diverted from other possible investments or uses, then the interest rate is the rate-of-return that could have otherwise been received for investing that money (e.g., spending the capital to construct the plant versus purchasing a bond).

Present Worth
Present worth (also called present value) is the current value of a future value. The future payments that make up a cash flow are discounted to reflect the time value of money. Present worth can be calculated to determine the effect of interest paid, the discount rate applied, or inflation. The mathematical definition of present worth is

$$PV = \frac{FV}{(1+i)^t}$$

where PV = present value of future monetary units t time periods in the future
i = discount or inflation rate per period
t = number of time periods

Table 8-2 represents the present worth of a simple series of future payments discounted at a rate of 5 percent. As shown, the present value of the future payments is less (discounted) each year.

Net Present Value

The NPV of a series of cash flows is simply the sum of the present worth of each of the anticipated cash flows:

$$NPV = (C_1 + C_2 + C_3 + C_4 + \cdots + C_n)$$

In the example present in Table 8-2, the NPV would be expressed as

$$NPV = (\$9,524 + \$9,070 + \$8,638 + \$8,277 + \$7,835) = \$43,344$$

So while the sum of the cash flow is $50,000, the NPV is $43,345, and given a 5 percent rate-of-return, $43,345 invested today is equivalent to receiving $10,000 per year for the next 5 years.

Escalation Rates

Escalation rate is that rate at which the costs of goods or services increase. Escalation rates typically considered in an LCC analysis are those of energy (purchased electricity and fuel), labor (operations and maintenance labor, as well as administrative labor costs), and parts and general maintenance costs. Escalation rates vary from country to country, by region, by industry, by labor source, and most importantly over time. The appropriate escalation rates this year may be much different from a decade ago or from 10 years from now.

Length of Analysis

The length of the LCC analysis is also an important consideration. A typical analysis length is approximately 20 years, though it will vary from project to project. Different portions of the project investment will have different useful service lives, and therefore the useful service life of the assembled project is difficult to estimate and use as a basis for the length of analysis. For example, the plant building may have a 50-year life, the generation equipment a 20-year life, and the piping a 30-year life. The further into the future the analysis looks, the harder it is to predict analysis variables such as interest rates, discount rates, and escalation rates, and the confidence level in the analysis is

Time period	Payments ($)	Present worth ($)
End of year 1	10,000	9,524
End of year 2	10,000	9,070
End of year 3	10,000	8,638
End of year 4	10,000	8,277
End of year 5	10,000	7,835
Total	0,000	43,344

TABLE 8-2   Present Worth Example (@ 5-percent Discount Rate)

therefore reduced. Project investors may also have a standard analysis length that is used for all analyses, or a specified time period in which they need to realize savings. Also, the farther out into the future values are, the more those values are discounted and, hence, the less those future values impact the NPV (e.g., at a 5-percent discount rate, a value in 20 years contributes 38 percent of its value toward the NPV versus a value in 30 years that only contributes 23 percent of its value toward the NPV).

Salvage Value

Salvage value is the value of the equipment, building, and so on at the end of its useful service life (or at the end of the economic analysis period). Whether or not salvage value is considered is primarily dependent on the length of the analysis, as discussed in the above paragraph. The salvage value of any equipment that reaches the end of its useful service life and is replaced during the economic analysis period should be considered in the cash flow. The actual effective amount of the salvage value and its significance to the overall analysis are affected by many factors, such as whether the equipment was fully depreciated at the point that it is taken out of service, market value of the equipment (either resale or scrap), which in turn can be affected strongly by the advance of related new technologies, and the cost of demolition, removal, or disposal of the item.

Equivalent Uniform Annualized Cost

Equivalent uniform annualized cost (EUAC) is a constant annual amount that is equivalent to the sum of all of the cash flows of an alternative. EUAC can be a useful metric in comparing alternatives with different project cash flows. The EUAC is the constant annual amount that could be paid and invested at the discount rate, which would cover the cash flow of payments. During initial periods (e.g., early years), the EUAC may be more than is needed to make annual payments, with the extra money invested at the discount rate, and in later years of the economical analysis, the EUAC may be less than is needed to make annual payments, with the extra money withdrawn from the previous years' savings.

Calculating LCC

Calculating LCC brings together the project cost, the annual operating and maintenance costs including energy costs, the cost of obtaining financing, any taxes, the time-value of money, and escalation into one value (e.g., the NPV) so that alternatives can be compared on a "like-for-like" basis.

The tables below provide a sample LCC calculation. Table 8-3 shows the economic factors assumed, while Table 8-4 shows the annual costs, escalated year-by-year over

Discount rate	5%
Maintenance escalation rate	3%
Natural gas escalation rate	2%
Electric escalation rate	2%
Administration and permitting cost escalation rate	3%

TABLE 8-3 Sample LCC Calculation Assumed Economic Factors

End of year	Annual Costs				Total annual costs ($)	Present worth ($)
	Maintenance cost ($)	Admin costs ($)	Fuel cost ($)	Purchased electricity cost ($)		
1	400,000	30,000	1,200,000	300,000	1,930,000	1,840,000
2	412,000	31,000	1,224,000	306,000	1,973,000	1,790,000
3	424,000	32,000	1,248,000	312,000	2,016,000	1,740,000
4	437,000	33,000	1,273,000	318,000	2,061,000	1,700,000
5	450,000	34,000	1,298,000	324,000	2,106,000	1,650,000
6	464,000	35,000	1,324,000	330,000	2,153,000	1,610,000
7	478,000	36,000	1,350,000	337,000	2,201,000	1,560,000
8	492,000	37,000	1,377,000	344,000	2,250,000	1,520,000
9	507,000	38,000	1,405,000	351,000	2,301,000	1,480,000
10	522,000	39,000	1,433,000	358,000	2,352,000	1,440,000
11	538,000	40,000	1,462,000	365,000	2,405,000	1,410,000
12	554,000	41,000	1,491,000	372,000	2,458,000	1,370,000
13	571,000	42,000	1,521,000	379,000	2,513,000	1,330,000
14	588,000	43,000	1,551,000	387,000	2,569,000	1,300,000
15	606,000	44,000	1,582,000	395,000	2,627,000	1,260,000
16	624,000	45,000	1,614,000	403,000	2,686,000	1,230,000
17	643,000	46,000	1,646,000	411,000	2,746,000	1,200,000
18	662,000	47,000	1,679,000	419,000	2,807,000	1,170,000
19	682,000	48,000	1,713,000	427,000	2,870,000	1,140,000
20	702,000	49,000	1,747,000	436,000	2,934,000	1,110,000
Total	10,756,000	790,000	29,138,000	7,274,000	47,958,000	28,850,000

TABLE 8-4   Sample LCC Calculation Annual Costs

Project cost	NPV of annual costs	Total present cost
$9,000,000	$28,850,000	$37,850,000

TABLE 8-5   Sample LCC Calculation Results

the project life and equated to the present worth, and Table 8-5 provides a resultant LCC calculation.

In the above sample calculation, the total present cost is $37,850,000, which is the sum of the $9,000,000 initial project cost and the $28,850,000 NPV of the annual costs. Note that the sum of the annual costs is nearly $48 million, but the future value is discounted to approximately $29 million. The total present costs of an alternative would then be compared with that of other alternatives to determine the alternative with the lowest LCC.

Determining Appropriate Escalation Rates

Various Federal Energy Management Building LCC (BLCC) Programs can be readily accessed via the Internet at http://www1.eere.energy.gov/femp/information/download_blcc.html#eerc and can be a great resource for determining appropriate energy escalation rates, including free downloadable software that calculates appropriate energy escalation rates individually and in a table format if needed. Some local public utility commissions also conduct fuel and electricity pricing forecast studies that are available to the public.

Additional (Secondary) Benefits and Value of Thermal Storage

There may be additional value created by the use of a thermal storage option. Some possible examples include

- Where chilled water storage may also act a fire protection reservoir, which will reduce risk and may even reduce insurance premiums
- Where thermal storage may provide additional redundancy and reliability above that provided by the BAU case
- Where thermal storage may provide useful operational flexibility beyond that provided by the BAU case.

Although it may be difficult to accurately calculate a dollar value for such secondary benefits, that difficulty should not mean that a value of zero is assigned (i.e., to ignore the value). Some attempt should be made to identify a quantitative (or at least a qualitative) benefit where applicable.

Making Recommendations

The final step in conducting a feasibility study is to develop recommendations based on the technical and economic analyses described earlier. As noted, typically, the recommended option is the option that is technically feasible and has the lowest LCC and/or highest NPV savings of the various alternatives considered; however, other factors, as noted, may affect the recommendation. Recommendations can/should also include any key observations that were made during the investigation process (e.g., add energy metering to measure usage, upgrade existing control systems, fix nonfunctional or poorly functioning systems/equipment). The recommendation should be included in a written report that summarizes the study findings as earlier shown in the sample table of contents.

A good feasibility study will provide the facility owner with a concept plan including thermal storage size and location, as well as an economic analysis including return on investment, and a good feasibility study can provide key data for obtaining utility rebates, future engineering design, and even commissioning efforts.

CHAPTER 9

Thermal Storage Design Applications

Introduction

This chapter describes a number of thermal storage design applications, discusses some specific needs associated with each type of application, and provides examples of actual applications. Since most of the thermal storage systems in use today use water or ice (or water with chemical additives), this chapter limits discussion to systems that use those thermal storage media.

There are and have been a wide variety of thermal storage system designs applied over the years; some have been successful and some have not. As might be expected, different thermal storage systems have somewhat different design considerations, and care must be exercised in the application of any thermal storage system to ensure that all design issues have been addressed.

Experience indicates that successful thermal storage design strategies typically employ the following:

- Simple system controls

- Single, constant, hydronic system pressure point

- High delta-T systems (note, not critical with ice harvester or external melt ice storage systems)

- Variable hydronic flow to meet the facility thermal load using variable frequency drive (VFD) pumps (also known as variable speed drive, or VSD) and two-way modulating control valves at the loads to achieve an as high, and constant, as possible, delta-T system (again, this is not always a consideration with ice storage systems that meet the load by melting ice in proportion to the actual load; nevertheless even those systems benefit from reduced pumping energy with variable flow applications)

- Top of water level in an open thermal storage tank is at least as high as the highest point in the hydronic system, if possible (note, while this is very desirable, if it is not feasible, methods are available, either directly or indirectly as discussed in this chapter, to account for coils or heat exchangers that are at higher elevations than the top of the atmospheric tank water level; however, as a rule, these fixes require more pumping energy than if the tank water level is at least as high as the highest point in the hydronic system).

As discussed in other chapters in this book, for a number of reasons, it is extremely important to maintain a constant and relatively large delta-T. Some of the reasons are smaller tank size for the same thermal capacity, smaller required distribution piping, and pumping energy savings. With sensible thermal storage, a constant delta-T is also necessary for proper operation of the thermal storage tank itself. The importance of high constant delta-T cannot be overstated as the key to the success with stratified water thermal storage. Consider, for example, that a chilled water (CHW) system operated or designed for 10°F versus 20°F delta-T requires twice the thermal storage tank size for the same ton-hour capacity and twice the CHW flow rate to supply the same facility cooling load. Therefore, in this example, the thermal storage tank costs are substantially higher, increasing initial system construction capital costs and reducing economic attractiveness (all else being equal). In addition, in theory, pumping energy is proportional to the cube of the flow. Thus, twice the flow rate requires eight times the pumping horsepower and resultant energy consumption. In practice, actual energy savings may be closer to the square of the flow (which is still substantial) due to fixed head losses, for example. In addition, when applied to existing systems, increasing the delta-T means that the existing piping systems can serve additional thermal load.

Of course, to help achieve high, relatively constant delta-Ts, VFD pumps are used to supply variable flow to serve facility thermal loads along with modulating two-way control valves at the loads to properly control flow. It is essential that the coil control valves be able to close and properly throttle against any pressure that may be produced by the pumping system. Note that larger delta-T can often be most efficiently produced by using thermal generation equipment in series. In addition, there are energy savings with series equipment, especially in high delta-T applications.

Alternatively, in lieu of pumping energy savings, distribution piping capital cost savings can be achieved with higher delta-T systems, as lower hydronic system flow is required to meet the thermal load, and thus smaller piping can be used for the same available pump head. Of course, most proposed or actual systems have a trade-off between some pump energy savings and some distribution pipe capital costs savings.

Thermal storage systems can be designed and operated with simple controls or with complex controls. It is no surprise that thermal storage systems with simple controls are easier to commission and maintain (especially for smaller or less sophisticated facility owners or operators). Figure 9-1 shows a basic flow diagram for a direct-connected CHW thermal storage system, with the thermal storage tank in the primary-secondary interface, where the water level in the thermal storage tank is higher than all of the hydronic system building coils or heat exchangers (or special provisions have been taken hydraulically to isolate those higher coils or heat exchangers). Note that none of the required isolation valves, pump appurtenances, or system controls is shown in this basic schematic. More importantly, also note that no control valves are required to control charging or discharging. With this simple thermal storage system, if the facility cooling load is less than chiller output, excess CHW (above the real-time facility cooling load) is stored in the thermal storage tank. Conversely, when the facility cooling load is greater than chiller output, CHW is discharged from the thermal storage tank to supplement the chiller output as needed to match the real-time facility cooling load. No special thermal storage charging or discharging controls are required per se (though chiller plant controls and other building system controls, of course, are still

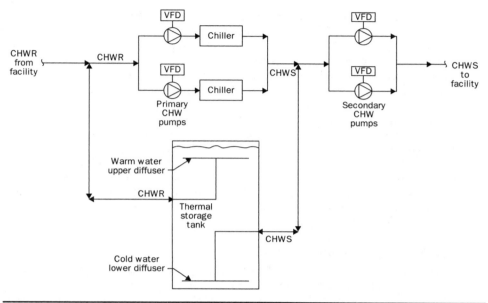

FIGURE 9-1 Thermal storage tank located in the primary-secondary interface.

required); and no complicated algorithms comparing temperatures and flows are needed. Chiller plant operator decisions, which can be automated, are limited to the number and timing of chiller operation to be dispatched. Note also that the thermal storage tank is located in a "bridge" between the chiller plant CHW supply (CHWS) and CHW return (CHWR) headers (the primary-secondary interface); it is important that no other bridge be open to flow, or CHW temperature quality can be adversely affected.

As a typical stratified water thermal storage tank is an open tank (i.e., vented to the atmosphere, not pressurized), it is preferred that the thermal storage tank be located so that the thermal storage tank water level is at least as high or higher than any other point in the hydronic system (i.e., higher than all the building(s) piping and coils or heat exchangers). Otherwise, pressure sustaining valves (PSV) or other pressure control or hydraulic separation methods (such as heat exchangers) are required to prevent possible overflowing the thermal storage tank and drawing air into the upper coils and piping.

With the thermal storage tank water level higher than all facility coils or heat exchangers, and the thermal storage tank located adjacent to thermal production equipment, the thermal storage tank becomes the hydronic system expansion tank and single constant pressure point in the hydronic system. Using the thermal storage tank as the expansion tank also means that there cannot be another open tank or expansion tank on the system since that would result in the ability to change the volume in a tank and change the location of the constant pressure point. The hydronic system itself undergoes varying dynamic pressures as piping distribution system pressure drop varies with the square of the flow and varies linearly with the distance from the pressure source, and every point can be calculated from the constant pressure point.

The preferred location of a thermal storage tank is often at the central plant. When a system is designed properly, locating the thermal storage tank at the central plant minimizes piping, minimizes the number of pumps and control valves required, and also greatly simplifies system controls. However, in some cases, thermal storage tanks must be located remotely from the central plant (e.g., there is no physical room at an existing central plant, or the central plant is located at a low point with respect to building coils and a higher elevation is available). If the distance is not too far (too far as determined by the construction cost to implement), not more than a few blocks, for example, the thermal storage system can still be configured as shown in Fig. 9-1. However, truly remote thermal storage tank interfaces are designed differently and are discussed in the following sections in this chapter. A remotely located tank can sometimes be desirable, in that it can be sited to act as a virtual satellite chiller plant during storage discharge, and thus can enhance the peak load carrying capacity of a CHW piping network (e.g., in cases of campus CHW systems where the network can be supplied from opposing points).

Caution Flags (Lessons Learned)

In contrast to successful thermal storage design strategies, but dealing with similar issues and challenges, there are a few caution flags that, if present, must be addressed carefully in the design of a successful thermal storage system. Not only must the design and construction address those critical "caution flag" issues, but the commissioning agent must also verify that the solution is working properly. The best thermal storage system in the world will not be entirely successful unless the thermal production plants, which provide the capacity, and the thermal systems, which are served, work together properly to meet the design objectives. Special care should be taken by all stakeholders to address the issues if any of the following conditions exist:

- Building coils or heat exchangers higher than the atmospheric (nonpressurized) thermal storage tank water level
- Multiple stratified thermal storage tanks on an open system (no heat exchanger between the tank and piping system)
- Lack of, or poor, delta-T control (a condition that should be fixed)
- High CHW distribution system differential pressures
- Systems with multiple pumps and valves, and complex controls
- Remote thermal storage tanks

Experience shows that high hydronic system differential pressures (delta-P) can contribute to low delta-T syndrome, as two-way control valves near the central plant can be forced open, thereby resulting in unused (no heat gain), bypass flow. That is, the two-way control valves cannot withstand the high delta-Ps, thereby causing excessive flows through coils or heat exchangers and resultant low delta-T due to the supply water flowing to the return header without controlled heat gain (or loss for heating systems). Therefore, as noted, it is essential that the two-way control valve be able to shut off and throttle against any pressure produced by the pumping system (use the centrifugal pump shut-off head as a basis for valve selection). Pressure independent control valves can greatly help with this challenge. Pressure differential control valves have also been used effectively by reducing and regulating differential pressure at coils or buildings close to the central plant to a differential pressure at which the control valves can function properly.

Multiple stratified thermal storage tank design (which is different from multitank design described in Chap. 4 and in this chapter) should proceed with special care. When adding a second stratified thermal storage tank, for example, flow cannot be simply doubled in the first, original thermal storage tank to charge or discharge the first thermal storage tank in half the time, since the diffusers were most likely not designed for the higher flow rate. As stratified thermal storage tanks are typically open tanks (at atmospheric pressure), water levels must be identical (or otherwise properly controlled) in each tank. If attempting to charge or discharge multiple thermal storage tanks simultaneously, hydraulic profiles for each thermal storage tank will need to be identical, unless an individual pumping arrangement is used. Extra care must be paid to hydraulic profiles (which are dynamic) and to the points of connection of the thermal storage tank to the hydronic system, to prevent surging in the thermal storage tanks.

While a complex control system can be made to work, it is nearly always better to keep it simple, if possible. Some thermal storage systems have multiple dedicated thermal storage pumps, multiple valves for various charging and discharging modes, and multiple temperature sensors for flow control. At one particular site, it was shown that the thermal storage system worked perfectly without the installed dedicated thermal storage pumps and valves.

Sensible Water Thermal Storage Systems

The most common water thermal storage systems in use today are stratified thermal storage, which includes both heating and cooling thermal storage systems. One primary difference between heating thermal storage and cooling thermal storage is that, with heating thermal storage, the supply water side of a heating storage system is connected to the top diffuser of the thermal storage tank and the return water side is connected to the bottom diffuser of the heating thermal storage tank. In cooling thermal storage, the connections are reversed compared to a heating thermal storage system. A second primary difference between heating thermal storage and cooling thermal storage, as discussed in other chapters, is the greater difference in buoyancy; with heating water systems, the much warmer heating water has a greater density difference per degree of temperature change than has the colder CHW systems. Therefore, greater care must be given with cooling stratified thermal storage than with heating thermal storage. A third difference between heating thermal storage and cooling thermal storage is the importance of insulation. The temperature difference between the storage media and the ambient air or ground temperature is much greater with heating thermal storage systems than with cooling thermal storage systems, which means that more insulation is required with heating storage systems. In addition to insulation of the thermal storage tank, an open tank system may need a floating insulation on top of the hot water to reduce heat losses to the atmosphere. Additionally, cold storage insulation systems should include a vapor barrier, which is not a requirement for hot storage systems.

Another difference relates to corrosion, as gases come out of solution as water is heated. Therefore, heating water systems have a much larger potential for air binding and corrosion as the water media is heated. For that reason, air venting and corrosion coatings are much more important with heating thermal storage systems than with cooling thermal storage systems (but are still important). Steps need to be taken to remove gases as close to the heating process as practical. In addition, hot water creates other corrosion problems and storage tanks need to be coated well, and a good corrosion inhibiter used in the water treatment system.

Water storage systems (both heating and cooling) can be divided into single and multiple tank systems. In addition, there are important differences between above ground and belowground thermal storage systems.

Single Tank Systems

Single tank stratified systems are the most common thermal storage systems in use today. One simple, successful flow or equipment arrangement is where the thermal storage tank and its piping form an interface between a primary and secondary loop as shown in Fig. 9-1. Water chillers and the thermal storage tank form the primary loop, and the secondary system to the building coils or heat exchangers form the secondary loop. As is typical with primary-secondary piping systems, either loop may circulate water independently of the other loop. There can be flow at any time, for example, from the warm CHW return at the top of the tank, through the chillers, lowering the water temperature, and then returning the CHW supply to the bottom of the thermal storage tank, thus charging the thermal storage tank. The building or coil loop circulates CHW supply from the bottom of the thermal storage tank through the buildings and coils and returns water to the top of the thermal storage tank. The chillers can operate at any time the CHW temperature in the storage tank (starting at the top of the tank) is at return storage temperature. The building or coil circuit can supply cooling at any time there is stored CHW in the thermal storage tank (starting at the bottom of the tank) or the chillers are operating. A heating system is similar except that solar heat, heat recovery, boilers, or some other heat source replaces the chillers and the diffuser connections are reversed, as previously described.

Because of the high cost necessary for a pressure vessel of great size, nearly all large water thermal storage systems are open tanks operating at atmospheric pressure (note, the tank is covered with a roof but is vented to atmosphere). There are several important implications of using an open, atmospheric tank, as discussed in the chapter introduction. One implication is that the thermal storage tank provides the piping system constant pressure point. Note that the thermal storage tank is the constant pressure point because there is constant atmospheric pressure at the top of the tank and because the water system is noncompressible. Since the water system is noncompressible, the volume remains the same, and since the action of pumps and piping losses does not change the volume, the elevation of the water in the tank remains the same (other than very small changes due to thermal expansion or contraction during discharging or charging). With the thermal storage tank as the constant pressure point, every other point in the piping system will relate to this atmospheric pressure point and differs in pressure by elevation and dynamic losses or gains.

Without pumps running, the static gauge pressure at any point in the hydronic system is equal to the elevation of the top of the water in the tank minus the elevation at the given point in the system. Similarly, in dynamic flow conditions, the gauge pressure at any point in the system is equal to the static pressure minus piping friction loss due to flow. For this reason, elevation of the top of the thermal storage tank water in relationship to coils is a key element. Also, this factor means that expansion and contraction of the water as it is heated or cooled occurs as a difference of elevation in the thermal storage tank, which in turn means that makeup water or overflow are controlled by the water elevation in the thermal storage tank. An operating deadband allowance will normally preclude the need for undesired makeup or overflow, except in unusual circumstances.

The hydraulics of an application should be carefully considered and additional controls may be needed in the actual application of Fig. 9-1. As an example, when the chillers are off and the distribution system is running with the intent of circulation flow through the storage tank, there will likely be undesired flow through the chillers unless automatic isolation valves are installed to prevent parasitic flow. In general, the flow will be inversely proportional to the resistance in each circuit, but any flow through nonoperating chillers would blend up the CHW supply temperature over that available from the thermal storage tank. Similarly, as discussed, if the coil valves are not closed, the operating chiller pumps can cause an undesirable parasitic flow in the CHW circuit to the facility without the distribution pumps running, which will result in some CHW bypassing the storage tank and lowering the inlet temperature to the chillers with the end result being a reduction in available CHW storage capacity. Either assuring that coil control valves are off when space cooling systems are not used or using flow isolation valves on the pumps eliminates this problem.

Thermal storage tank connections

Chilled and hot water thermal storage tanks may be either direct connected or indirect connected. Direct connected means that there is no heat exchanger between the thermal storage tank and the facility chilled or hot water system. Figure 9-1 is an example of a direct connected thermal storage system. Indirect connected means that the thermal storage system is not directly connected to the facility hydronic system, but instead has a heat exchanger interface, which hydraulically separates the thermal storage tank from the facility hydronic system. Direct connected systems may be as shown in Fig. 9-1, or may have pumps, PSV, and two-way control valves, as shown in Fig. 9-2.

Thermal storage systems connected as shown in Fig. 9-1 have the advantage of requiring fewer pumps, no thermal storage control valves, simpler plant system controls than other direct and indirect systems, and the advantage of not having the pressure drop loss (i.e., the flow energy loss) across the PSV. However, a thermal storage system

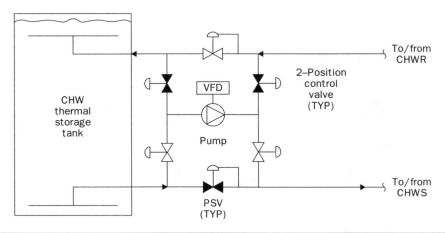

FIGURE 9-2 Direct connection using modulating PSV (shown in discharge mode).

as shown in Fig. 9-1, as discussed, must be located at, or very close to, the thermal production equipment (e.g., located at or next to a central plant using primary-secondary CHW pumping), and the water level in the tank must be higher than all of the facility or building coils or heat exchangers, unless special measures have been provided for any specific cases where building coils or heat exchangers are higher than the thermal storage tank water level. With the system shown in Fig. 9-2, in operation, the thermal storage tank, when charging, acts as an additional cooling load on the distribution system, and, when discharging, acts similar to a chiller. It should be noted that this causes substantial differences in pressure differential between the supply and return piping mains depending on whether the system is charging or discharging, and is most pronounced near the thermal storage tank.

Where a specific building or coil(s) is higher than the top of the proposed thermal storage tank water level, that building or coil can be isolated through the use of intermediate heat exchangers, or through the use of building PSV and building or coil pumps. At some point, when there are too many buildings or coils or heat exchangers above the top of the proposed thermal storage tank water level (e.g., when installing a relatively short or below ground thermal storage tank), then it may be more cost-effective to connect the thermal storage tank as shown in Fig. 9-2 or Fig. 9-3 (i.e., to take care of the water height difference at one location versus multiple locations).

Returning to Fig. 9-1, the actual number of chillers can be as many as best suited, even though only two chillers are shown for this example. Likewise, the number of pumps can be as many as best suited, even though only two pumps are shown for the primary pumps and two pumps are shown for the secondary pumps. In addition, CHW pumps can be dedicated to individual chillers, as shown in this example, or can be installed with common suction and discharge headers, which can allow for a backup pump(s) to be installed more affordably. The VFDs on the primary CHW pumps can be used for balancing to maintain a constant primary loop flow rate through the chillers, or the VFDs can modulate the primary CHW pumps to keep the chillers fully loaded with less than design delta-Ts, or the VFD CHW pumps can be used as part of an overall all-variable speed system. The secondary CHW pumps modulate to maintain remote minimum system differential pressure or a most open or fully open control valve strategy (e.g., the VFD decreases secondary CHW pump speed until one control valve

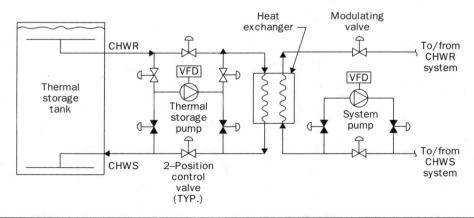

FIGURE 9-3   Indirect connection using a heat exchanger (charging mode shown).

out in the system is 100 percent open and then speeds up a little bit until the particular control valve starts to close, which indicates that the modulating valve is in control).

Figure 9-2 shows a direct connection CHW thermal storage system using modulating PSV for the case where building coils or heat exchanger are higher than the top of the thermal storage tank water level. Note that as in Fig. 9-1, this Fig. 9-2 does not show all of the valves, fittings, pump appurtenances, and controls required, and likewise, multiple pumps can be used as best suited, even though only one pump is shown for this example. Figure 9-2 shows the CHW thermal storage tank operating in the discharge mode with the thermal storage pump drawing CHW supply from the bottom of the thermal storage tank and pumping into the facility CHW supply system, and with CHW return returning to the top of the thermal storage tank through the PSV. The PSV maintains positive pressure (e.g., 5 psig) at the highest coil or heat exchanger elevation by throttling as necessary (note, the thermal storage PSV can receive a signal from that remote coil location or an equivalent signal directly from the local sensing point as indicated in the PSV symbol used in Fig. 9-2). In the charging mode, the valves shown as closed (solid colored), will be open, and the valves shown as open, will be closed; the thermal storage pump draws warm water from the top of the thermal storage tank and pumps it into the CHW return system to the system chillers, where it can be cooled and pumped back to the CHW thermal storage tank through the PSV, which operates identically to the other PSV just described. When using PSV, care must be exercised to avoid slamming the valve shut as flow is cut off, which can cause hydraulic shock and subsequent damage to the system, in some cases. At the same time, the design must assure that enough pressure is maintained to prevent vacuum from occurring in the high parts of the system, which can draw in air through valve stems, air vents, or other parts of the system, and may even cause damage to the coils if high vacuum conditions are created. Note that one factor affecting this issue is to maintain pressure on the supply side as well as the return side, which may mean a check valve in the supply to close if the CHW supply in the circuit is cut off.

In lieu of using PSV, water-turbine-driven pumps have been used, which use the required pressure drop of the system to drive the thermal storage pump, with the help of some additional electric power. This type of system can be very energy efficient compared to the use of PSV; however, proposed energy savings need to be weighed against additional capital costs (note, the best applications are where the flows are very large and the pressure differential is high).

In some cases, so much of the CHW (or hot water) system is above the thermal storage tank water and the consequences of failure are so great that heat exchangers are used to isolate the thermal storage tank as shown in Fig. 9-3. A heat exchanger offers a positive way to control pressure in the distribution system while using an open tank. The use of heat exchangers (and the extra associated pump set), however, is expensive and also requires a larger thermal storage tank than direct connection to a thermal storage tank because of the difference in (or "approach") temperature between the two sides of the heat exchanger. The heat exchangers ordinarily used for this application are of plate-and-frame type to obtain the closest approach temperature technically and economically feasible. Even so, the CHW supply in the storage tank must be somewhat warmer than the CHW supplied by the chillers to the heat exchanger, and the return water to the chillers is not as warm as the return water in the storage tank. These factors combine to mean that isolating heat exchangers are not often used with CHW even though they are often used with ice storage systems.

As with the other figures in this chapter, isolation valves, controls, and appurtenances are not shown in Fig. 9-3, and additional heat exchangers may be used as best suited, even though only one heat exchanger is shown. Figure 9-3 shows the system in the charging mode with relatively warm tank CHWR drawn from the top of the thermal storage tank and pumped through the heat exchanger, where the thermal storage tank water is cooled by the system CHW supply to become tank CHWS, and pumped back to the bottom of the thermal storage tank. In the charging mode, the thermal storage pump speed can be modulated with a VFD to maintain the heat exchanger leaving CHWS temperature back to the thermal storage tank. In the charging mode, the system CHWS flow, which is provided by central plant chillers and pumps, can be modulated with a two-way control valve to maintain system differential pressure or a specific flow rate, for example. Note that the system pump , shown in the figure, does not operate in the charging mode, assuming that the central plant distribution pumps are able to circulate flow through the heat exchanger system.

In the discharge mode, the two-position valves on the thermal storage side of the heat exchanger, shown in Fig. 9-3, reverse position (i.e., open valves close, and closed valves open) so that the cold tank CHWS is drawn from the bottom of the thermal storage tank and is pumped through the heat exchanger in the reverse direction from the charging mode, where the tank CHWS is warmed, as it cools the system CHWR, to become tank CHWR, and back to the top of the thermal storage tank. In the discharge mode, the thermal storage pump speed can be modulated with a VFD to maintain the heat exchanger leaving system CHWS temperature supplied to the facility CHWS system. In the discharge mode, on the system side, the modulating two-way valve on the CHWR is opened fully (or a separate bypass, not shown, is opened) and the two-position control valve positions are reversed from that shown in Fig. 9-3, so that the system pump draws system CHWR through the heat exchanger and pumps into the facility CHWS system in order to meet or help meet facility cooling demand.

Note that the flow reverses on both sides of the heat exchanger from the charge mode to the discharge mode, and that the heat exchanger flow is always counter flow. If a particular heat exchanger manufacturer or model does not allow reverse flow, a valve arrangement can allow for unidirectional flow so that the heat exchanger inlet is supplied with either CHWS or CHWR.

The thermal storage system connections shown in Figs. 9-2 and 9-3 can also be used for remote thermal storage locations (remote from the central plant), where the thermal storage tank acts as a satellite plant during discharging and as a load during charging. In this case, in the charging mode, the PSV on the CHWS pipe shown in Fig. 9-2 modulates to maintain pressure, and the thermal storage pumps maintain CHWR pressure at the central plant. If the top of the thermal storage tank water level is above building coils or heat exchangers, and individual cases with higher coils or heat exchanger have been addressed, the PSV can be replaced with two-way valves (the CHWS valve is still modulating).

Note that, in all cases, the thermal storage pump always pumps out of the thermal storage tank.

Multiple Tank Systems

Multiple tank systems pose additional considerations. With more than one open tank, if there is any dynamic pressure difference in the piping to each tank, there will be water added to one tank and taken from another. A fairly small difference in pressure head (e.g., a foot or two of water column) would result in water removed from one tank possibly overflowing to the other tank. In a series tank arrangement, the operating head is different in each tank. The most common series tanks are closed pressurized tanks; however, often the pressure in these types of thermal storage tanks is relatively low. In some cases, higher than expected pressure leads to tank damage.

As an example, consider three closed storage tanks in series. Typically, the middle tank would be connected to an open expansion tank with the elevation of the water in that tank below the maximum operating pressure for the tanks. In operation, the middle tank forms the constant pressure point and one tank has slightly higher pressure and one tank a slightly lower pressure.

A second arrangement for multiple tanks is tanks in parallel. To work properly, the tanks must connect to piping systems at the same point and must have identical dynamic pressure losses. Any differences in dynamic pressure losses would cause a different operating water level in the tanks.

A possible third solution is to provide controls similar to those used on swimming pool surge tanks and to install water level controls on at least one tank (usually the smaller diameter tank) to maintain a design water level, which means that there must also be some different way to control expansion and makeup water (e.g., using level controls on the second, larger diameter tank). The fourth way is to install plate-and-frame exchangers between one or both of the tanks and the hydronic system.

Diffuser Design

With a stratified water thermal storage system, the purpose of the upper and lower diffusers is to introduce and withdraw the water from the tank in a laminar fashion, in order to minimize mixing between the warm water at the top and the cold water at the bottom of the thermal storage tank. With a good diffuser design, the thermocline thickness should be less than 3-feet (ft) thick. There are different types of diffuser designs, which have been used successfully including radial-flow (plate or disk) diffusers and slotted or perforated pipe diffusers. Other diffuser solutions have also been designed and applied. As an example, an early chilled water design at Stanford University used a distribution header in a bed of gravel to provide a uniform discharge and intake for the CHWS over the entire area of the thermal storage tank. No matter the design, all good diffuser designs are based on obtaining appropriately low Froude and Reynolds numbers (the Froude number is the ratio of the inertial forces to the gravitational forces, and the Reynolds number is the ratio of inertial forces to viscous forces). Froude numbers should typically be less than 1.0 for good diffuser design, although higher Froude numbers have been reportedly used successfully. The maximum allowable Reynolds number is, in part, a function of the thermal storage tank water height, with higher allowable Reynolds number for taller thermal storage tanks (45 ft and taller). However, a lower Reynolds number is usually better, and, per the ASHRAE handbook, tests show that there is no mixing due to inertial forces below a Reynolds number of approximately 250 and negligible mixing below a Reynolds number of approximately 450. As a maximum allowable, it is suggested that the Reynolds number should not exceed 2000 for taller tanks (at least for piping diffuser systems). Other factors, such as tank geometry, can affect thermocline thickness and subsequent thermal storage tank performance.

Per ASHRAE Cool Storage Design Guide, the inlet Froude number, Fr, is defined as

$$Fr = \frac{Q}{(gh^3(\Delta\rho/\rho))^{0.5}}$$

where Q = volume flow rate per unit length of diffuser, ft³/s·ft
  g = gravitational acceleration, ft/s²
  h = inlet opening height, ft
  ρ = inlet water density, lb/ft³
  $\Delta\rho$ = difference in density between cold and warm water in the tank, lb/ft³.

Note that ft³/s·ft is cubic-feet-per-second-per-foot, ft/s² is feet-per-second squared, and lb/ft³ is pound-mass-per-cubic foot.
The inlet Reynolds number (Re) is defined as follows:

$$\text{Re} = \frac{Q}{v}$$

where v = kinematic viscosity, ft²/s (square-feet-per-second).
Note that the Reynolds number above should be divided by 2 for arc slot diffusers, since the velocity vector has two directions.
Note that since the maximum flow rate is typically given and since the CHWS and CHWR temperatures are often set by an existing facilities infrastructure, the only two variables that can actually vary are the flow per unit length and the diffuser inlet opening height. The flow per unit length can be decreased to obtain a lower Froude number by increasing the diffuser length or area; and similarly, the height of the diffuser inlet, above the floor for the bottom diffuser and below the water level for the upper diffuser, can be increased to obtain a lower Froude number. Similarly, the Reynolds number, in this case, is solely a function of diffuser length, or of diffuser area (perimeter circumference and depth) for a radial disk diffuser, for a given maximum flow rate and known water viscosity, which is a function of temperature (use the average viscosity).
Prior to beginning the diffuser design, the thermal storage tank size should be determined as discussed in Chap. 7 so that the required thermal tank parameters are known. Key information or parameters to be supplied by the owner for the diffuser design includes

- Thermal storage tank thermal capacity
- Thermal storage tank dimensions
- Peak flow rate
- Supply and return temperatures
- A drawing of the proposed tank including any interior column dimensions

Given the above information, the first step in slotted piping diffuser design is to layout the diffusers, typically using a single or double octagon (although triple or quadruple octagons or other shapes are possible, such as an H-shape, often used for rectangular tanks), depending on the peak flow rate, the diameter of the proposed thermal storage tank, and the subsequent Froude and Reynolds number calculations. Based on the "ASHRAE Design Guide for Cool Thermal Storage," ideally, with a double octagon diffuser, the interior diffuser serves about 50 percent of the tank plan area and outer diffuser serves about 50 percent of the tank plan area. With a basic layout determined, the length of diffuser piping can be measured. Given the total peak-flow rate through the thermal storage tank and the estimated length of diffuser piping, the flow per unit length can easily be calculated by dividing the two numbers. The Reynolds number can also be calculated once the flow rate per unit length is known. Where the slots are located, diffuser piping diameter should be sized for a flow velocity of less than 1 ft/s so that there is negligible pressure drop between different points in the slotted diffuser piping. Additionally, diffuser supply piping should be symmetrical as shown in Fig. 9-4. Given the thermal storage supply and return temperatures, the water densities can be determined, and with the flow per unit length already determined, the diffuser height

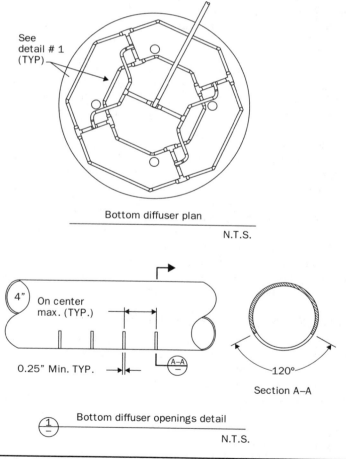

FIGURE 9-4 Sample pipe diffuser plan and section (job specific).

can be set in order to achieve a Froude number of 1.0 or less using the equation previously given.

The slot width and number of slots is determined by assuming a uniform flow velocity exiting the slots, not to exceed 1 ft/s. Given the total peak flow volume through the thermal storage tank and a maximum 1 ft/s slot flow velocity, the total flow area can be determined (flow volume equals flow velocity multiplied by flow area). Given the length of diffuser piping, and assuming equal slot spacing, such that slot spacing is not more than twice the diffuser height (i.e., the number used in the Froude calculation), the number of slot openings can be determined. Given the number of slot openings, the total required slot flow area, the diameter of the slotted diffuser pipe, and using 120° slots, the minimum required slot width to achieve the required area can be calculated. In addition to the area of the slots, the design of the slot through the pipe affects the discharge velocity and mixing at the discharge of the slot. A good diffuser design needs to have enough losses to maintain equal discharge in all the slots (i.e., a relatively high delta-P across each slot relative to the delta-P along the inside of the slotted piping) while providing a very low discharge velocity into the tank.

Given a maximum allowable diffuser pressure drop (10 ft of water column, for example, which includes both the upper and lower diffusers as well as interior tank piping) and the maximum 1 ft/s flow velocity discussed, the diffuser supply piping can be sized such that the calculated pressure drop is less than the maximum allowable.

The above procedure is provided as an example of designing a diffuser. However, note that a very large segment of the stratified tank market is served by suppliers of "turnkey" thermal storage tanks, who take full responsibility for the design, supply, and thermal performance of the thermal storage tank, including the flow diffusers.

Also, note that most slotted piping diffusers have been constructed using plastic (PVC) piping. There reportedly have been a number of documented structural failures in the past of plastic pipe diffuser installations. The failures have been attributed to failures during system pressure spikes (water hammer) as well as to failures during tank filling or draining (when piping is filled or emptied of water, more slowly than the surrounding tank volume, causing pipe hanger or pipe failure). Accordingly, provisions or care must be made to allow for proper drainage or venting during draining or filling operations, pipe supports should account for buoyancy, and large valves must not be allowed to slam closed; stand pipes can be designed to prevent the translation of water hammer. Additionally, some engineers or designers specify that any piping internal to the storage tank must be welded steel piping.

Instrumentation and Control Requirements

The key instrumentation on a stratified thermal storage tank comprises the array of temperature sensors that measure the temperature of the water at increments of elevation from near the bottom of the tank to near the top of the tank water level. The space between sensors is a balance between cost and providing enough data to monitor operation of the storage tank. Often temperature sensors are installed in intervals of approximately every 2 ft vertically. On a tall tank this gives a fairly good idea of the stored capacity and depth of the thermocline. However, with a shorter tank, more of the capacity will be in each vertical foot of tank height and a smaller spacing is needed. The thermal storage tank temperature sensors allow facility operators and control systems to determine or monitor the charge of the thermal storage tank. The tank temperature sensors can also be used to help determine the thickness of the thermocline. On a steel thermal storage tank, the temperature sensors, set in thermowells, are typically located vertically along the side of the tank, next to the ladder or stairway for ease of access; on concrete (and some steel) tanks, the temperature sensors are often hung in a bundle from a small roof opening.

Additional thermal storage instrumentation includes two temperature sensors at the tank inlet/outlet piping (one each in the CHWS and CHWR piping), and a bidirectional flow sensor, on either the supply pipe or the return pipe, in order to calculate thermal capacity into and out of the thermal storage tank.

Finally, the thermal storage tank should have standard tank level instrumentation including low-level alarm, open-fill valve, close-fill valve, high-level alarm, and overflow alarm. A water meter is also often put on the make-up fill line.

Additional Design Requirements

If the thermal storage tank is a design-build effort, the engineer may also be required to determine insulation requirements or thicknesses and to conduct heat gain or loss calculations to show that the estimated heat gain or loss is less than the maximum allowable (e.g., no more than 2 percent of rated thermal capacity per day). Heat gain or loss calculations are performed based on the principles and formulas from the physics of heat transfer. The majority of heat gain or loss is typically via conduction, which is proportional to the exposed (above ground) tank surface area; the temperature difference between the outside air and the fluid inside the tank; the overall heat transfer coefficient, which is a function of the type and thickness of tank construction materials, as well as the wind velocity; ground temperature; floor and soil conductivity; solar insolation on the exposed tank; and losses through nozzles, fittings, ladder supports, and so on are also contributing factors.

Chemically Modified CHW Storage Systems

There are proprietary chemical additives, which lower the freezing point and maximum density point of water to create a low temperature fluid (LTF), for operation of stratified sensible heat storage below the 39.4°F temperature of maximum density of plain water. Most of the applications to date have been on relatively large CHW systems. While the systems that use those additives are sensible storage systems and operate essentially the same as CHW storage systems, there are some differences that affect design consideration.

The chemicals are used to increase the amount of storage that can be provided by a given volume of water. Conversely, a 50-percent increase in delta-T is a 33-percent reduction in tank size for a given amount of thermal storage. The chemicals are expensive (which may roughly offset the cost savings from the reduced tank volume), but the systems offer some other benefits in addition to tank size. In application, these systems may be considered almost a hybrid of ice and water storage as they offer some of the benefits and shortcomings of each of those systems. Since the LTF systems are sensible storage, assuming that series chillers are employed, not all the stored cooling is obtained at the same temperature point, which means that for the upstream chillers the cooling is obtained at greater efficiency than at the final temperature obtained by the final chiller. Colder CHW can be obtained with this solution than with standard CHW and may lead to smaller distribution piping, reduced pumping energy, and use of the existing distribution system to serve more load than was originally designed. While there is a slight energy penalty with this solution, the energy penalty is, typically, not as great as with ice systems, which provide all their thermal storage at a single low point (well below the subsequent storage discharge temperature) and often use large pumped volumes to freeze ice.

These LTF systems usually employ plate-and-frame heat exchangers to minimize the amount of chemical needed. In operation, the LTF systems are similar to other plate-and-frame solutions used on CHW, except that the chillers are usually piped directly to the tank so that the coldest coolant produced is stored in the tank and the antifreeze characteristics of the solution protect the chillers. Some systems pipe the LTF solution directly to the loads; although this requires more expense for the additives, it eliminates the need for a plate-and-frame heat exchanger, provides the benefits of the larger delta-T to the distribution pumps and piping and to the load heat exchangers, and can provide antifreeze protection to cooling coils in freezing weather. In fact, some applications, which open CHW coil valves in freezing weather and produce CHW to serve special loads, instead of operating a chiller at that time, would benefit from the freeze protection provided by the LTF.

When using colder than design CHW, whether from CHW storage or a chemically modified system, it may be desirable to install blend up controls at building or coils because coils are designed for a certain degree temperature rise. Providing colder water increases the capacity of the coil, but when controls modulate the control valves closed, the coil capacity is then reduced by the lower velocity water flow in the coil. The consequence is that while colder water is supplied, there may not be much greater rise in the coil than the original design, if the flow becomes laminar, which can mean a system designed for 45°F supply and 60°F return when supplied with 35°F CHW only produces 50°F CHW return. But if the supply water at the coil is blended up to 45°F, it can obtain the 60°F return. That increases the delta-T on the thermal storage tank, chillers, and distribution piping from 15° to 25° with a corresponding increase in storage capacity and an even greater reduction in pump energy.

An additional significant benefit from at least one commercially applied LTF is its inherent ability to control both corrosion and microbiological activity, precluding the need for any on-going water treatment for those functions.

Ice Storage System Design

All of the different ice storage technologies, understandably, have unique system design features and requirements. However, the ice-on-coil, internal melt system provides a reasonable illustration of many of the issues common to the different storage technologies. Figure 9-5 represents a very simple, but typical, cooling storage design, with both the charging (night ice-making) and discharging (day cooling) modes depicted.

The chiller and thermal storage are located in a primary loop, usually with a constant flow rate, although variable primary systems are beginning to appear. A separate, variable flow pump delivers coolant to the secondary cooling load circuit. In the case represented, the chiller and storage are in a series flow configuration with the chiller as the first component encountered by the coolant returning from the secondary loop (chiller upstream). Other arrangements are possible, including reversing the position of the chiller and storage or placing them in parallel.

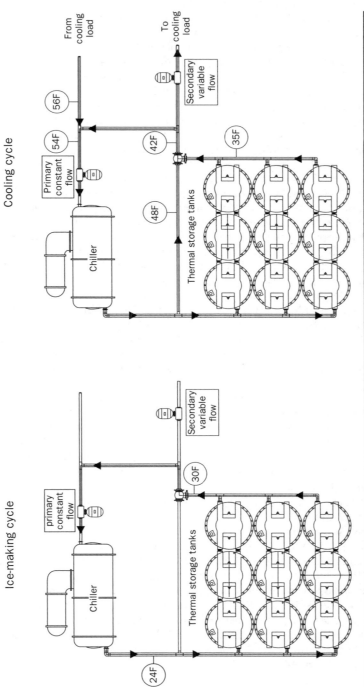

FIGURE 9-5 Charging and discharging modes for an internal melt ice storage system utilizing a secondary coolant.

173

Because the temperature of the coolant leaving the thermal storage tanks varies with the quantity of ice in inventory, the storage inlet temperature, and the coolant flow rate, a three-way modulating valve is often employed to automatically adjust the volume of coolant that flows through, or bypasses, the storage tanks.

During the ice-making mode, cold coolant flows from the chiller, then passing through the storage tank heat exchangers. The coolant rises in temperature as it removes heat from the storage tank water, exiting the tank at slightly less than 32°F and returning to the chiller. Coolant temperatures gradually decrease as the ice accumulates, and, as discussed in the description of internal melt equipment, the coolant temperature drops much more rapidly when most of the water is frozen and the latent heat is exhausted. The relatively abrupt change in the rate of temperature decrease can be used as an indication that the ice-making function is complete and should be terminated. Another indicator of a "full charge" sometimes utilized is the rise in the water level that accompanies the change in density as water is frozen.

After the ice inventory is completely restored, the chillers are turned off and the ice-making mode is not reactivated until the next ice-making period, usually the following evening. When refrigerating machinery operates at ice-making conditions, its capacity is reduced to approximately 65 to 70 percent of its nominal rating when operating at conventional CHW temperatures. The impact on refrigeration or chiller equipment efficiency depends on a variety of factors, including ambient temperature depression at night, type of ice storage, type of chiller, and so on.

During the subsequent cooling period, a series arrangement simply illustrates the flexibility provided by ice storage systems. Depending on the original design intent, there may be periods when all of the cooling load can be met entirely by the chiller, or by the ice storage system, or by a combination of the chiller and ice storage in different proportions. A common design feature of series ice storage designs is the increased difference between the system supply and return temperatures over the traditional 10 to 12°F (5 to 7°C). For partial storage systems, the smaller chiller, often only half the capacity of chillers in conventional systems without thermal storage, may not be able to pass the flow that corresponds to peak load with the smaller delta-T. By increasing the temperature difference, often to a range of 14 to 16°F (8 to 9°C), the coolant flow is reduced significantly for the same load. The system depicted in Fig. 9-5 uses a 14°F delta-T. The lower temperatures available from ice systems provide the opportunity to reduce the supply temperature, expanding the coolant temperature differential.

In many cases, it is desirable that the chillers operate at maximum capacity before any storage is depleted. This is easily accomplished by setting the chiller leaving-chilled-water-temperature (LCWT) at the design system supply temperature, in the example case, 42°F. As long as the chiller is capable of supplying this temperature of coolant, the three-way valve will bypass all of the flow around the storage tanks. When the chiller is no longer capable of meeting the entire cooling load, or it is preferable to shift cooling load to the storage system, its leaving coolant temperature will rise above the desired design supply value. The three-way modulating valve, controlled by the downstream coolant temperature, will automatically redirect some of the coolant through the storage tank so that the blended flow downstream of the valve will be at the desired temperature. The flow through the ice storage will vary over the course of the day. As the ice is melted, or as the load increases, more flow will be diverted through the ice storage tanks.

An intermediate storage condition, and a cooling load less than the design peak, is also represented in Fig. 9-5. The storage system is capable of providing 35°F, still well below the design limit of 42°F. The three-way valve only needs to divert a portion of the 48°F flow through the storage tanks to achieve the desired 42°F delivered to the cooling load. In a perfectly designed system, all of the flow would be directed through storage, exiting at the target temperature of 42°F, at the worst-case combination of cooling load and ice inventory. In other words, the storage capacity is selected so that the design supply temperature is assured for all cooling loads up to and including the peak design load, at any condition of ice inventory during which those loads are expected to occur.

Since the snapshot of the system occurs during a time when loads are not at peak design, not all of the primary flow is delivered to the secondary cooling loop. Some of it bypasses, to be mixed with the flow returning from the cooling loads. The mixed coolant, now at a temperature of 53°F flows through the chiller, is cooled to 48°F, and the circuit is complete. Note that the chiller in a series ice storage system, located upstream of the storage tanks, almost always operates at a higher, and more efficient, leaving temperature than it would in a conventional system without thermal storage during the on-peak period.

More of the cooling load can be shifted to the storage system by simply increasing the chiller leaving temperature setpoint or demand limiting the chiller. Even for partial storage systems, there may likely be times of the year when on-peak chiller operation can be avoided completely, and manipulation of the chiller or storage temperature is a simple and effective means to adjust the relative contributions of the two components.

The same control goals can be achieved with the storage and chiller locations reversed. In this case, the leaving chiller temperature would be held constant and the storage temperature, again controlled by a three-way valve, will determine the chiller inlet temperature and thereby the relative contributions of the chiller and storage to the cooling load. For instance, if the storage leaving temperature is set to the design supply temperature, the chiller can be completely idle or the valve can direct all flow to bypass storage, shifting the entire load to the chiller.

Larger ice storage systems will often consist of many storage tanks, sometimes in the hundreds, with tens of thousands of ton-hours storage capacity. These systems will often have multiple chillers, sometimes in parallel with the storage tanks, but the control objectives are unchanged.

Expanding an Existing Thermal Storage System

Perhaps the most common regret of those who install a large central thermal storage system is that they wish they had installed a larger system. That is, facility owners are very pleased with the installed thermal storage system, but when confronted with expanded thermal loads (e.g., due to new buildings) facility owners are faced with how best to expand the existing system. This follow-on facility expansion can be challenging and much more expensive than it would have been to install a larger thermal storage tank initially. The cost per unit of thermal storage tank capacity drops with an increase in tank size. However, that savings, obviously, cannot be obtained after the initial thermal storage tank installation. So, given an option, a project should take a close look at future loads and at whether the funds can be obtained to build the larger thermal storage system initially to accommodate the future loads. Of course, nearly every project has budget constraints, and a project usually has to be cost-effective based on the thermal loads served when the project is installed. However, that means that very commonly a central facility has to deal with the fact that later they have an undersized thermal storage system for the expanded thermal load, or that additional energy savings could be obtained with a larger installed thermal storage system.

Expanding an existing thermal storage system is usually not a simple matter. In the first place, an open tank system, where the CHW or hot water is directly supplied to a building or group of buildings, cannot easily accommodate an additional thermal storage tank; that is due to the almost inescapable probability of hydraulic imbalance between the thermal storage tanks, resulting in overflow in one tank while the other tank loses water. A second factor is often that the desired expansion is small in comparison to the installed capacity.

There are, however, several ways that existing thermal storage capacity can be expanded. Most common methods to expand existing thermal storage capacity are using multiple tanks hydraulically balanced; using multiple tanks with plate-and-frame exchangers isolating storage from the distribution system; increasing the existing thermal storage tank height; modification of existing coils, circulation piping, and chillers to increase delta-T; using a chemical additive to increase the delta-T; conversion to ice storage; and using hybrid chilled water-ice systems.

Multiple Tanks Hydraulically and Statically Balanced

In some cases, a second thermal storage tank can be added with the same size, elevation, connection point to the hydronic system, and identical dynamic losses as the original thermal storage tank. In this case, the static and dynamic pressures must be balanced and the water level must remain the same in both tanks in both a static and dynamic situation. From a use point-of-view, adding a second thermal storage tank can be equivalent to doubling the tank capacity, and having the two-tank system function as a single tank.

Multiple Tanks with Plate-and-Frame Exchangers

The drawback of having multiple points at atmospheric pressure that is associated with more than one point of connection is the pressure imbalance that can occur between thermal storage tanks, even if the tanks have the same water level when not in use. If a pressure imbalance occurs, water will flow into the thermal storage tank with the lower pressure connection and out of the other tank with the higher pressure connection, which can be a difficult problem to overcome (though it has been accomplished using level controls). One way of dealing with this problem is to install plate-and-frame heat exchangers on one or both thermal storage tanks. This means that, for the hydronic circulation system, it no longer has more than one atmospheric tank. The designer must also install an expansion tank with this modification if all tanks are isolated by heat exchangers. The addition of heat exchangers will require pumps and control valves, and the amount of thermal storage will be reduced because the delta-T will be lower as previously discussed.

Increasing the Thermal Storage Tank Height

Another way to increase capacity of an existing thermal storage system, without the added pumps, control valves, loss of delta-T, and problem of multiple points at atmospheric pressure, is to increase the height of an existing thermal storage tank (note, this solution only applies to steel tanks). The added tank wall sections must be added to the bottom of the thermal storage tank because the greater tank height will mean greater hydraulic pressure at the bottom of the tank, which may exceed the working pressure of the original thermal storage tank design. To accomplish this modification the thermal storage system must be out of use during the modification. The process of increasing an existing thermal storage tank height involves disconnecting the tank (and the upper flow diffuser) from the base and floor and raising the tank with a crane, with jacks, or by some other method. The original tank must be suspended during the installation of new lower sections or else be set off to the side. The new lower sections are set in place and welded in a fashion similar to that of the original thermal storage tank. Once the new sections of the tank are in place, the original tank can be set on top of the new section(s) and welded into place. After testing, coating, and insulating it, the modified thermal storage tank operates in the manner as if a taller tank was originally designed and installed.

Modification of Coils, Circulation Piping Controls, and Chillers

Thermal storage systems are often installed on existing systems that use older engineering practices in coil design and delta-T. As an example, chillers may have a small delta-T such as 10 to 16°F and that limitation is used in the entire system design. In fact, many older systems have 45°F CHW supply temperature and 55 to 58°F CHW return temperature. When coils use that low a rise in temperature (i.e., low delta-T), it sometimes may be difficult to obtain a greater rise even when supplied with colder CHW. Often a simple fix to obtaining greater delta-T is a pump on the coils and controls to blend up colder CHW (say 40°F) to the design CHW temperature. So far as coil performance is concerned, there is no change; but an additional 5° in the storage and circulation system is obtained. If a system initially has a 15° rise, adding an additional 5° to storage means a 33-percent increase in thermal storage capacity. This idea of increasing delta-T can be expanded by replacing existing coils with those designed for much higher delta-T. Changes in coil design can not only handle colder CHW supply, but at the same time provide a higher CHWR temperature. A system with coils designed for 45°F CHWS temperature and 58°F CHWR temperature when replaced with coils with a 40°F CHWS temperature and 65°F CHWR temperature provides an increase in the delta-T from 13 to 25°F or nearly double the capacity of the thermal storage tank (as well as correspondingly lower circulation volumes and pumping horsepower). On a system with only a few large coils, replacing coils offers a very inexpensive way to increase thermal storage capacity.

A common concern when discussing coil change out is a possible increase in fan horsepower on air-handling systems. However, increased fan horsepower does not have to be the case as the primary difference in coil design is often a change in the number of circuits. Any increase in pressure drop is usually on the water side; however, there are compensating reductions in hydraulic losses due to reduced CHW flow.

It is also possible in some cases to increase the delta-T across existing coils, simply by lowering the CHWS temperature (and often also achieving a simultaneous increase in CHWR temperature), if the proper flow control valves are in place at the coils.

Chemical Additive to Increase the Delta-T

As discussed in other sections, the CHW supply temperature in a CHW storage system is restricted by the maximum density point for water. This point is about 39.4°F. This means that as the temperature of CHW drops toward 39°F, it becomes denser (heavier) and tends to settle toward the bottom of a storage device. But below 39°F, plain water becomes lighter (less dense) and tends to rise. This means CHW storage tank systems cannot be designed to use much lower than 39°F CHW without the addition of chemical additives.

Moreover, as previously discussed, chemical additives are available to modify the maximum density point so that maximum density occurs at a temperature lower than 39.4°F. However, there are other factors to consider when changing to the use of chemical additives. For one, as noted, the LTFs are relatively expensive, which needs to be considered when compared with other solutions. Second, chiller energy consumption and costs are somewhat higher to produce 32°F LTF than to produce 40°F CHW; so a modified system will not be quite as energy efficient. On the other hand, colder allowable CHW, which will allow for a larger delta-T, can provide greater thermal storage capacity for an existing thermal storage tank, allow for less pumping horse power, and allow existing piping mains to serve greater connected loads. Note that existing chillers may or may not be able to produce the colder LTF needed to take advantage of this solution; but, in many cases of large chillers, modifications (if needed) are possible to allow the chillers to produce colder water (or new low temperature chillers can be added).

In addition, other differences between plain water and the LTF must be considered. First the LTF has a higher density and weighs more than plain water, which may limit the use (or maximum operating depth) of the LTF in an existing thermal storage tank. There are also differences in specific heat and other factors that affect flow and horsepower. And, due to costs and other factors, it is usually desirable to isolate the tank and chemical solution from the CHW circulation system, which requires a plate-and-frame exchanger to isolate the solution from the distribution CHW.

Conversion to Ice Storage

In some cases, a CHW system has been converted to an ice thermal storage system. Two different conversion solutions have been used. In some cases, an ice harvester has been added on top of the existing thermal storage tank with the ice dropping into the tank, which adds substantial capacity to the tank but is a major modification affecting almost every element of the system design. In particular, it is not possible to use the chillers to provide thermal storage when the water in the tank is at a lower temperature than the chiller's leaving CHW temperature. This solution also mixes the temperature in the thermal storage tank; so the thermal storage tank no longer functions as stratified storage.

A second solution is the installation of ice-making coils (or encapsulated ice containers) in the water thermal storage tank. As far as the function of the storage tank, this functions in a similar way to that with ice harvesters; however, the type of ice production equipment is different. In some cases, the existing chillers can be modified to produce cold brine for use in the ice-making coils (or around the encapsulated ice containers).

These conversions are in a sense dual or hybrid systems. In most cases, this means the ice is used for thermal storage while conventional chillers provide CHW for immediate use. In the case of full storage, the ice storage is used only during the on-peak period and the chillers are held off at that time. In the case of partial thermal storage, chillers provide CHW for immediate use and the ice storage is used to add capacity to that of the chiller during peak loads.

Hybrid Chilled Water-Ice Systems

Hybrid chiller water-ice thermal storage systems differ from those described above in that the CHW storage system is retained and an ice storage system is added. These hybrid thermal storage systems function as separate thermal storage systems, and thermal storage can be provided by each system in a method that would be used if each were the only component. When supplying CHW for use, the ice thermal storage system is piped to a plate-and-frame heat exchanger in series with the CHW discharge from the CHW storage tank. This hybrid system can be controlled for a greater delta-T all the time, or the ice component can be held off until loads exceed those desired to be supplied by the CHW thermal storage tank. While it typically takes more energy to produce the ice than CHW, the greater system delta-T provides pump energy savings, serves more capacity, and is a relatively simple fix to adding more thermal storage capacity. During lower load periods, it is possible to hold the ice thermal storage system off and obtain the greater savings possible with the original CHW thermal storage system.

Ultimately, the best choice for adding capacity is project specific and depends on many unique factors, including the magnitude of the desired increment of capacity to be added to the original thermal storage installation, as well as the estimated cost to implement any option.

CHAPTER **10**

Thermal Storage Tank Specifications

Introduction

When using large stratified thermal storage, the thermal storage tank specification is one of the key specifications among the project technical specifications. This chapter provides an overview of both steel and concrete thermal storage tank specifications. (The reader should realize that, as tank construction requirements vary by type of tank and by the project requirements themselves, virtually any project specifications are job specific and must be edited for the job.) Other key project technical specifications include other large equipment specifications, controls (including sequence of operations), test and balancing, and commissioning, as well as Division 1 specifications.

Most construction specifications, whether for thermal storage systems or for any other type of equipment or material, are divided into three main parts: Part 1—General; Part 2—Products/Material; and Part 3—Execution/Construction. Part 1—General of the specification specifies: a basic description of the specified system and scope of work; submittal requirements; design standards, codes, and references; design requirements; warranties and penalties; and quality assurance requirements. Part 2—Products outlines or delineates specific requirements for that item, material, or product being specified. Part 3—Execution specifies construction and start-up requirements, which in the case of aboveground steel tanks, for example, include the following: tank welding and inspection; the application and testing or inspection of coatings; the application of insulation; hydrostatic testing for leaks; performance tests; operation and maintenance manual requirements; guarantees or warranties; and terms of final acceptance.

A thermal storage specification and other documents (such as drawings, geotechnical report, etc.) define the requirements for construction standards, materials, fabrication, and required tank appurtenances for tank, foundation, and flow diffusers. Aboveground thermal storage tanks are constructed from steel or concrete. Steel thermal storage tanks must be built aboveground (unless installed in a silo) and are often constructed in accordance with American Water Works Association (AWWA) Standard D100 "Welded Carbon Steel Tanks for Water Storage." Partially buried and belowground tanks are constructed from concrete, typically, in accordance with different AWWA Standards (e.g., D110 or D115), although other materials besides concrete can be used, especially for shallow multiple tank designs. Additionally, the majority of thermal storage tanks specifications are written as turnkey specifications with the thermal storage tank builder

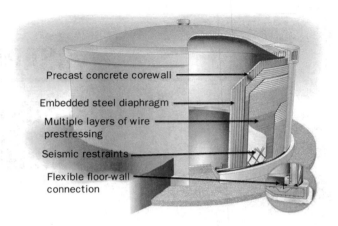

Precast concrete corewall

Embedded steel diaphragm

Multiple layers of wire prestressing

Seismic restraints

Flexible floor-wall connection

FIGURE 10-1 Concrete thermal storage tank (courtesy of Natgun).

responsible for the thermal performance of the tank (i.e., meeting the specified thermal storage capacity); the maximum allowable heat loss over the specified period; the maximum allowable pressure drop across the thermal storage tank inlet and outlet nozzles based on the specified maximum water flow rate; the structural design of the thermal storage tank and foundation; and for any required soil remediation or augmentation.

Aboveground Steel Thermal Storage Tank Specifications

This section provides and discusses a typical aboveground, steel, water thermal storage tank specification. As noted, the thermal storage tank specification is typically divided into three parts, although sometimes more parts (e.g., Part 4) can be or are used.

Part 1—Execution

The thermal storage specification should typically begin with a basic discussion of the project or system including, for example, the following: the specified thermal storage capacity; that the water storage tank should function as part of a chilled (or hot) water thermal storage system; that the storage tank should be used for meeting cooling (or heating) thermal loads, and that thermal storage tank should be designed and constructed to ensure proper thermal stratification for storing relatively warm and cold water based on specified temperatures. Assuming a turnkey thermal storage tank project, the specification should also explicitly state that the thermal storage tank contractor should be responsible for the design, manufacture, and construction of all aspects of the thermal storage tank including foundation and soil remediation or augmentation (if necessary), the internal flow diffusers, as well as the thermal performance guarantee or warranty of the storage tank, as specified. The basic description should also highlight the basic operating strategy, such as, for example, "the thermal storage tank shall be integrated directly with the existing chilled water supply system, and shall store chilled water produced during 'off-peak' periods for use during the following facility 'on-peak' cooling loads."

The basic description also outlines the expected scope of work, such as, for example, "the work shall include, but is not limited to, designing and providing the following thermal storage tank items: one aboveground stratified thermal storage tank with fixed roof; tank foundation and associated earthworks; tank anchorage; tank fittings and accessories, including ladder, railing, clips for attachment of instrumentation conduits, and other elements shown on drawings and described in this specification; flow diffusers; tank coatings; roof and wall thermal insulation; and provide (or make provision for, depending on how the work is divided) instrumentation." Note, sometimes, projects assign the thermal storage tank instrumentation to the tank contractor, and, sometimes, projects assign the thermal storage tank instrumentation to the project controls contractor.

Finally, the basic description should also state that the geotechnical report (sometimes called a "soils report") is part of the thermal storage project construction contract documents and must be provided by the facility owner to all the bidding tank contractors with the bidding documents.

Submittals

Part 1 also specifies the submittal requirements. Submittals are documents produced by the contractor for review by the engineer and owner in order to help verify compliance with the specifications and contract documents, and must be "approved" by the engineer (i.e., "no exception taken" or "make corrections noted"), before the contractor may purchase any materials or proceed with the work, except at his or her sole risk. Typical, aboveground, steel thermal storage tank submittal requirements include having the contractor provide the following nine items:

1. The following items certified, signed, and sealed by a professional civil or structural engineer registered in _____ (fill in the local jurisdiction), with at least five (5) years (as a minimum) previous experience preparing TES tank drawings:
    a. Drawings indicating tank orientation and elevations, all information pertaining to diameter, height, plate thickness, and grades
    b. Fabrication and erection drawings including the interface
    c. Drawings and specifications of the foundation and the excavation and backfill underneath the tank
    d. Drawings of all of the tank penetrations and points of interface
    e. Structural design calculations for the tank, anchor bolts, and foundation system; these calculations should include all assumptions and design loads
    f. Details of all weld joints
2. Internal tank elevation indicating water level with respect to internal sensors, flow diffusers, and other components
3. The following product data:
    a. Coating data sheets and color samples for owner's review and approval
    b. Insulation:
        i. Details and data sheets. Include roof and shell details, roof-to-shell and shell-to-base transition details, and penetration details
        ii. Heat loss and gain calculations, confirming compliance with this specification, including all tank surfaces and including an allowance,

based on experience, with thermal storage tank construction for losses through appurtenances such as ladders, railings, manholes, and vents

4. Performance calculations:
   a. Thermal storage calculations demonstrating compliance with the requirements of this specification
   b. Pressure drop calculation for diffusers and internal piping demonstrating compliance with this specification
5. Submit the records, reports, and other documentation specified throughout this thermal storage tank specification
6. Test Results as described in this specification
7. Layout drawings and written procedure indicating phasing and coordination of work with all related design drawings and existing conditions
8. Radiographic film and inspection reports from certified welding inspector
9. Welders' qualification records and certifications

Design Standards, Codes, and References

Another key portion of Part 1 specifies the design standards, codes, and references for steel thermal storage tank construction and may include the following:

- American Concrete Institute (ACI)
  - ACI 301—Specification for Structural Concrete
  - ACI 318—Building Code Requirements for Reinforced Concrete
- American Society of Heating, Refrigeration, and Air Conditioning Engineers (ASHRAE)
  - ASHRAE HVAC Systems and Equipment, Chap. 50, Thermal Storage
  - ASHRAE Design Guide for Cool Thermal Storage
- American Society of Mechanical Engineers (ASME)
  - ASME Section IX—Welding Procedures
  - ASME B31.1—Code for Pressure Piping–Power Piping
- American Water Works Association (AWWA)
  - AWWA D100—Welded Steel Tanks for Water Storage
  - AWWA D102—Coating Steel Water Storage Tanks
- Air-conditioning & Refrigeration Institute (ARI)
  - Guideline T—Specifying the Thermal Performance of Cool Storage Equipment
  - Standard 900, Thermal Storage Equipment Used for Cooling
- Steel Structures Painting Council (SSPC)
  - Manual and SSPC Standards, Volume 1 and 2
- State and local code of regulations or building codes

Thermal Storage Tank Design Requirements

Part 1 of the specifications should further describe the thermal storage basic design requirements. Typical thermal storage basic design requirements for an aboveground, steel, thermal storage tank include the following:

A. Thermal and flow requirements (six items)
  1. General design requirements (the tank specifier or purchaser should define these item-1 requirements):
     a. Tank capacity—Note the thermal capacity of the thermocline, the tank air space (ullage), and expansion volume; the water below the bottom flow diffuser and the water above the upper flow diffuser should not be considered as useful thermal capacity when calculating or demonstrating the tank capacity or performance
     b. Hydronic system supply and return temperatures sent to the tank during charging and discharging, which also provides the delta-T
     c. Maximum flow rate to and from the thermal storage tank for both charging and discharging conditions
     d. Definition of the thermocline, which is defined as "the band of water separating the cooler water at the bottom of the tank from the warmer water at the top of the tank"; the bottom surface of the thermocline is (often) defined as water of a temperature 2°F above the cold water temperature supplied to the tank during charging, and the top surface of the thermocline is defined as water of a temperature 2°F below the warm water temperature returned to the tank during discharging
  2. The tank design-build contractor shall demonstrate the thermal storage capacity both mathematically and by means of a performance test, as detailed in this specification; the calculation should assume that the full volume of the thermocline is present in the tank at the fully charged and fully discharged conditions

  3. Recharging of the thermal storage tank shall occur primarily during the off-peak electrical rate period; the design of the tank (for a chilled water system) shall allow the following operation: warm water shall be removed from the top of the tank at up to the maximum design flow rate; cold water will be added at an equal flow rate to the bottom of the tank; recharging may continue until full cool storage capacity is attained or may stop when the tank is only partially charged; the design of the tank shall allow the charge rate to vary from a minimum of zero gallons per minute (gpm) to the maximum design flow rate without affecting the performance of the tank or reducing the rated storage capacity

  4. Discharging of the thermal storage tank shall occur to meet facility load requirements; the design of the tank (for a chilled water system) shall allow the following operation: cold water from the bottom of the thermal storage tank shall be removed at up to the maximum design flow rate; warm water shall be returned to the top of the tank at an equal flow rate; discharge may continue until the entire thermal storage capacity has been delivered or may stop when the tank is only partially discharged; the design of the tank shall allow the discharge rate to vary from a minimum of zero gpm to the

maximum specified design flow rate without affecting the performance of the tank or reducing the rated storage capacity

5. The thermal storage tank should* limit the ambient heat gain (or heat loss, for a hot water thermal storage system) to less than 2 percent of the rated thermal storage capacity in a 24-hour period based on a __ °F dry-bulb outside air temperature (fill in the local maximum design temperature), maximum solar heat gain, and a fully charged water temperature of __ °F (fill in the minimum chilled water supply temperature for cooling applications)

6. The total hydraulic head loss from the tank inlet flange to the tank outlet flange at the maximum design flow rate should be a maximum of 10 feet (ft) of water column [3 pounds per square inch difference (psid)]

B. The thermal storage tank contractor should design, provide, and test the tank, foundations, and appurtenances in accordance with the latest edition of AWWA D100 and local building codes, as well as any requirement necessary to receive certification from any "authority having jurisdiction"; the tank, foundations and appurtenances should be designed to withstand the predicted ground settlements but not less than specified in the soils report; the thermal storage tank specification should specify the following five design requirements:

1. Wind load: each component of the tank, anchors, and foundations should be designed for the most stringent wind load specified in AWWA D100 and the local building code

2. The tank, anchorage, foundation, and appurtenances should be designed per AWWA D100 and the local building code; seismic loads should be determined in accordance with the following criteria:
   a. For building code design: Site coefficient SD, seismic importance factor
   b. For AWWA D100 design: Site Class, Seismic Use Group
   c. Seismic loads associated with vertical ground acceleration should also be included

3. Roof live load: minimum roof design

4. Corrosion allowance

5. Anchor the tank properly to the foundation; foundation anchor bolts should be used whether or not they are required under the documents listed in this specification (note, anchorage may possibly not be used if there is no uplift during any design condition):
   a. If anchor bolts are required by the documents listed in this specification, then install as required by these documents
   b. If anchor bolts are not required by the documents listed in this specification, then install anyway, using the minimum size and maximum spacing allowed under AWWA D100, Sec. 3.8

---

* note that in some cases, as in this example sample or chapter specification, the author has used the word "should" to indicate some consideration, freedom of choice, adjustment by the reader; however, actual specifications should use the word "shall" to indicate a mandatory requirement

C. Calculate the required expansion volume based on a piping system volume of ___ gallons (fill in the hydronic system volume), not including the thermal storage tank, a tank volume that meets the requirements of the specification, a minimum system temperature of __°F, and a maximum system temperature of __°F; the thermal storage tank shall include room for expansion of the entire calculated chilled water system volume; indicate the minimum working water level and the maximum working water level on the tank submittal

D. Tank height: as recommended to accommodate the water level that meets the thermal requirements of the specification, seismic sloshing wave, expansion, freeboard, and roof peaking requirements but shall not exceed ___ feet maximum (fill in the maximum height allowed for the project, if there is no height limit, delete the last clause); include freeboard calculated in accordance with the formula in AWWA D100 13.5.4.4; contractor should submit diagrams to demonstrate conformance with these requirements, indicating relative water level; the minimum distance from the maximum working water level to the bottom of the overflow should be ___ inches (in.) (typically, at least 2 in, depending on the volume-to-height ratio of the tank); the minimum working water level should be at least 6 in. above the upper distribution header (or diffuser plate), or higher if recommended by the tank manufacturer

E. The contractor shall design and provide provisions for mounting control and instrumentation conduit on ladder supports without damaging insulation vapor barrier and metal jacket

F. Earthworks
  1. The preparation of the soil beneath the tank and tank foundation must be in accordance with the project geotechnical report; the contractor should, as a minimum, over-excavate and re-compact, the soil underneath the tank per the requirements of the project geotechnical report (minimum 3-ft overexcavation); refer to specification section "Excavating and Backfilling for Structures (a separate project specification, not discussed here)"; the contractor shall accommodate the properties of this soil in the design of the tank and foundation, per the requirements of the geotechnical report; backfill must be placed also in accordance with the requirements of the geotechnical report
  2. The contractor should design and provide a foundation meeting the requirements of AWWA D100 and ACI 318; prior to commencing the work, the contractor should submit to owner's representative a method statement for the backfilling operations

Part 1 also specifies the warranty and thermal penalty requirements. Typical warranty period is one year from substantial completion; however, virtually any warranty length can be specified, keeping in mind that the facility owner will pay (perhaps significantly) for extended warranties. One way the thermal performance penalty can be calculated is to divide the thermal storage project cost by the thermal storage capacity and assess that unit cost (or greater if circumstances warrant) to a thermal storage capacity shortfall. For example, if through performance testing, the commissioning agent determined that the thermal storage capacity was 1000 ton-hours less than a specified minimum of 10,000 ton-hours, the contractor could be assessed a back charge of 0.0001 times the contract sum for each ton-hour of thermal storage capacity not met (1000 ton-hours equal to 10 percent of the contract amount, in this case).

Finally, Part 1 typically also specifies the minimum required guarantee or warranty.

Part 2—Products

Part 2—Products specify the specific requirements, standards, materials for the item specified, and, in the case of an aboveground, steel, stratified thermal storage tank, Part 2 typically specifies the following:

- Acceptable thermal storage tank construction contractors
- Tank construction itself
- Tank foundation, as applicable
- Tank coatings
- Tank insulation
- Internal flow diffusers
- Tank fittings and accessories

Each of the above items is covered under a separate section, often called paragraphs, for example, Par. 2.1 Tank Construction. The following provides a description of the specific requirements for an aboveground steel tank, the most common type of large-scale cooling thermal storage tanks.

Part 2 usually begins with a list of acceptable manufacturers or specialty contractors able to provide a turnkey thermal storage tank. The tank contractors should have a minimum number of years in business and a minimum number (e.g., five) of similar projects. Tank construction should be specified per the requirements of the construction standard used for the project, such as AWWA D100 or API 650 and should fill in any blanks and specify any options required by the standard. Each standard provides a checklist of required items that the owner needs to provide to the tank construction contractor. As a minimum, the specification paragraph on basic tank construction should note the type of tank, that the storage tank shall be an aboveground, vertical, cylindrical, flat bottom, fixed roof tank of all-welded steel construction and shall be ground-supported, for example. Further, that the bottom and roof steel plates of the tank shall be assembled using lap joints, welded continuously from the top side only, and that the shell plates are to be 100-percent full fusion butt-welded joints. Additionally, the roof steel should have a smooth, continuous slope without warpage to prevent the ponding of water, and the roof plates should be supported so that roof does not exhibit "oil canning."

Foundation

The foundation is a critical component of a large aboveground thermal storage tank, and must be specified properly, or, in a worst case, undue differential settlement may occur. The specifications should state that the thermal storage tank contractor must design and provide a foundation, meeting the requirements of the latest versions of AWWA D100 and ACI 318 (or a similar standard), and must adhere to the project geotechnical report, which must be included as part of the construction contract documents. Where the recommendations of the geotechnical report are less stringent than the requirements of AWWA D100 or the local building code, the tank contractor should use the more stringent requirements. The exposed surface of the foundation must be sloped away from the tank to prevent the accumulation of rainwater. The tank should be shimmed and grouted with a sand-cement mixture.

Concrete The concrete parameters of the foundation must be specified and reference made to the separate project concrete specification(s) (e.g., Section 03300). A basic example of concrete foundation parameters follows.

   A. Refer to Division 3—concrete specifications. Minimum acceptable concrete mix parameters are:
   1. Concrete compressive strength: 3000 psi [project dependent as determined by the structural engineer (SE)].
   2. Slump: 3 in. ± 1 in.
   3. Total air content: 3 percent.
   4. Maximum aggregate size: 1 to 1/2 in.
   5. Water-cement ratio of the concrete should not exceed 0.5 for non-air-entrained concrete.

The thermal storage tank should be constructed with a 6-in. minimum sand cushion supplied under the tank. The sand shall pass a 3/4-in. sieve with less than 5 percent retention on a 3/8-in. sieve, and less than 10 percent passing a #200 sieve. The sand should be clean, free from clay balls and lumps of earth and should be free of corrosive materials, such as salt and sulfur. The maximum chloride content should be less than 100 parts per million (ppm) and the maximum sulfate content should be less than 200 ppm. When a ring-wall foundation is used, the inside of the ring-wall should be insulated with a minimum 2.5-in. thickness of polystyrene board (or approved equivalent) extending 32 in. below the top of the ring-wall (or the maximum depth of the ring-wall if less than 32 in.).

Coatings

In order to minimize corrosion of steel surfaces, protective coatings can be applied to the thermal storage tank. The effectiveness of a coating depends upon a number of factors, including (maybe most) importantly the surface preparation, the type of coating used, the number and type of coatings (e.g., primer and top coat), the thickness applied, the curing or drying process used, as well as the type of fluid (including additives) to be used inside the tank. Typically, an epoxy material complying with the requirements of AWWA Standard D102, Internal Coatings, is used and should comply with the Volatile Organic Compound (VOC) content limits required by State or Federal legislation or guidance, as well as local air pollution control district (or air quality management district) requirements. The total dry film thickness of the epoxy coating should typically be a minimum of 8 mils (0.008 in.) and a maximum of 18 mils. No negative tolerance on thickness should be permitted.

The exterior of the tank should also be primed, and the weld surfaces ground and primed in accordance with AWWA D102, and any exposed areas epoxy coated with a typical minimum total dry film thickness of 4.5 mils and a maximum of 7.0 mils, per AWWA D102, or other appropriate standard. Color (for any exposed surfaces) should match insulation jacket. Under the insulation, one primer coat (5 mil) conforming to AWWA D102 should be applied. The paint system should be applied to tank prior to insulation. The underside of the bottom of the tank does not need to be coated. Any stainless steel, aluminum, galvanized, and nonmetallic surfaces require no additional coatings. Erection damage to galvanized surfaces, if any, should be touched up with zinc paint.

Insulation

Having the appropriate amount of insulation is another critical installed component of a successful thermal storage tank. If insulation is not sufficient, there can be excessive heat gain with cool storage (excessive heat loss with hot storage), while too much insulation wastes precious capital. Insulation can be specified to be provided by the tank contractor as a turnkey performance requirement, with a minimum wall and minimum roof insulation thickness. The insulation system should meet or exceed the system performance requirements for the specified heat gain (e.g., a maximum allowed heat gain in a 24-hour period is equal to 2 percent of thermal storage capacity). It is important that the insulation system be designed to withstand the maximum wind loads specified for the project location.

The entire tank roof and shell must be insulated and provided with a vapor barrier seal. The vapor barrier must be continuous with seams and joints between panels sealed. One insulation method is to use rigid insulation board, of polyurethane or polyisocyanurate, with pre-attached aluminum foil vapor barriers. The insulation board should typically have a minimum density of 2 pounds per cubic foot PCF and an aged thermal conductivity value no more than 0.18 Btu-in./hr-ft2-°F (i.e., an aged insulating value of at least, R = 5. 6/in.). The outermost layer of the insulation board should be bonded (laminated) to an aluminum jacket, of 0.03 in minimum thickness, to provide weather and mechanical protection. External banding or screw attachment of the aluminum jacket should be prohibited. The attachment system for the insulation panels should be a concealed fastener system such as tensioned cables and pencil rods. The exposed insulation jacket should have an embossed or textured finish (so minor dents or dings will not be so noticeable), with a factory painted surface of a color selected from vendor's standard colors by an appropriate owner facility representative.

The roof insulation (or a portion of the roof insulation) made needs to be walked on, and as such the tank design builder needs to design the insulation system to withstand the specified live load and to accommodate tank deflection under these loads (or else it needs to be understood and agreed that the roof insulation is not to be designed or used as a walking surface).

Additionally, the insulation system should include removable insulated covers for the following:

1. Manholes and access panels

2. Anchor bolt chairs

3. Mounting flanges for controls

4. All piping connections

5. All other penetrations to the roof or wall systems for removable equipment, instrumentation, or other items

Finally, with respect to insulation, the perimeter area, where the tank shell meets the foundation, should be adequately sealed and insulated to minimize condensation.

Flow Diffusers

Often, the flow diffusers are designed and provided as part of the turnkey (i.e., design build) thermal storage tank effort; however, the performance requirements must still be specified. For example, the thermal storage specification must state the turnkey thermal storage tank contractor shall design and provide the internal flow diffusers required to meet the thermal performance and thermal storage tank dimensions specified in the specification. The design should be based upon proven flow distribution calculations to maintain the operating thermocline between the cold and warm water, over the full range of flows and operation of the proposed thermal storage tank.

The thermal storage specification should state that the diffuser design should be per the "ASHRAE Design Guide for Cool Storage," Chap. 4 (or ASHRAE HVAC Systems and Equipment Handbook, Chap. 50). As discussed in Chap. 9, there are two basic types of flow diffusers: pipe (slotted or perforated) diffusers and radial flow (disk or plate) diffusers. A radial disk type diffuser should meet the following specific requirements:

- Maximum inlet Froude (Fr_i) number should be 0.6.
- Diffusers should be self-balancing diffusers within the storage tank to be symmetrical relative to the vertical axis of the tank and to the horizontal center plane.
- Diffuser should not exceed 50 percent of the plan area of the tank.
- Diffuser opening height should not exceed 5 percent of the overall depth of water in the tank.
- All piping components inside the tank should be welded carbon steel.
- The diffuser should be inherently robust in order to minimize damage from infrequent, unanticipated water-hammer events that could be initiated elsewhere in the hydraulic system apart from the thermal storage tank.

A pipe slot type diffuser should also meet the following additional modified and specific requirements:

- Froude number of less than 1.0 at outlet
- Reynolds Number less than 1000 (the appropriate value depends on depth of tank, with taller tanks being able to use higher Reynolds number, with minimal mixing due to fluid velocity below Reynolds number of approximately 450)
- Outlet flow velocity at 1 ft/s or less
- Pipe material—epoxy-coated steel, or schedule 80 PVC (note, some thermal storage tank builders also use schedule 40 PVC, and each pipe material choice has advantages and disadvantages)
- Pipe supports—stainless steel, or fiberglass, attached to floor and roof (note that any pipe hangers should accommodate uplift for events when the diffuser could be buoyant, e.g., during tank filling when the piping may be filled with air, while surrounded at least in part by water)

Tank Fittings and Accessories

The project plans should show the desired location for the various required tank fittings and accessories as described below. The thermal storage tank contractor should submit all fittings and accessories locations on shop drawings for engineer review. Typically, all piping connections should be located within approximately 3 ft of ground level. The thermal storage tank should have, as a minimum, the following 13 items covering fittings and accessories:

1. Two 36-in. diameter shell access manhole with hinged door, opening outward, near ground level; the bottom of the access manholes should be above the bottom of the drain

2. One 24-in. diameter and either one 36-in. diameter or one 36-in. × 36-in. square rainproof roof access hatches. Locate one hatch near the roof center and one near the roof edge; hatches should be fully hinged and self-supporting to allow unencumbered access

3. One warm water inlet/outlet nozzle, 150-psi flange; provide a nozzle size that does not degrade the specified pressure drop or thermal performance; the piping from the nozzle to the internal flow distributor should run inside the tank; the engineer should specify the minimum nozzle diameter

4. One chilled water inlet/outlet nozzle, 150-psi flange; provide a nozzle size that does not degrade the specified pressure drop or thermal performance; the piping from the nozzle to the internal flow distributor should run inside the tank. Minimum nozzle diameter is __ in. (fill in based on flow)

5. One overflow nozzle with an internal overflow pipe and an external seize-resistant, anti-thermosiphon device within 5 ft of grade; overflow to be sized for the maximum flow entering the tank through the makeup water connection at the rate of __ gpm (equal to the maximum fill rate); locate overflow discharge as shown in the drawings

6. One 6-in. diameter center column clean out nozzle and blind flange

7. One roof vent designed and constructed to prevent entrance of insects, birds, or animals; vent should have the capacity to pass air so that at the maximum possible rate of water, either entering or leaving the tank, excessive pressure shall not be developed; the overflow pipe shall not be considered a tank vent

8. One 4-in. diameter auxiliary drain nozzle with centerline located 2 ft above grade; include valve and threaded cap; inside tank, drain should elbow down to 3 in. above the tank bottom

9. Grounding lugs as specified in subparagraph 6-4.1.4 (c) of National Fire Protection Association (NFPA) 780

10. Controls and Instrumentation
    a. Provide welded carbon steel threaded thermowells located at 18-in. to 2-ft intervals of height, 18-in. insertion length, and ¾-in. diameter, mounted along the tank access ladder or staircase for ease of maintenance; pressure rating of well is to be consistent with the system pressure in which it is to be installed
    a. Provide tank-level control and alarm mounting locations as shown in the drawings; mounting locations should be easily accessible for control and instrumentation installation and maintenance

11. (Optional) One 6-in. diameter fire protection connection connecting to wye fitting with dual 4-in. NST-threaded fire hose connections; provide gate valve between tank and each fire-protection nozzle; gate valve should be AWWA C-509 resilient seated valve with outside stem and yoke (OS&Y) construction; provide locks on gate valves

12. Carbon Steel Butterfly valves on inlet and outlet nozzles: 150 pounds per square inch gauge (psig) cold working pressure (CWP) rating, tight shutoff, flanged type with one piece stem, 200°F rating, 316 stainless steel disc and shaft or bronze disc with 416 stainless steel shaft, PTFE-coated stainless bearing, EPDM shaft seal, EPDM seal material, with manual sintered bronze gear operator, and handwheel

13. Provide an external access caged ladder (or spiral staircase, at owner's request) with lockable access and meeting regulatory requirements

Part 3—Execution

Steel tanks can last for many decades if constructed (that is welded together), coated, and insulated properly. Coatings and painting must be applied properly to be effective and, if not properly applied, can actually lead to premature failure due to concentrated localized corrosion.

Tank Welding

All welders should be qualified by ASME Section IX, or another appropriate standard, for the processes and positions utilized. The contractor should employ the services of a welding supervisor independent of the tank erection foreman's jurisdiction.

The edges or surfaces of the pieces to be joined by welding should be prepared by flame cutting, plasma arc cutting, arc gouging, machining, shearing, grinding, or chipping and must be cleaned of detrimental oil, grease, scale, and rust.

Welding procedures should also be in accordance with ASME Section IX. All field welding may be done by the shielded metal arc welding process, the gas metal arc welding process, the flux core arc welding process, or the submerged arc welding process. Shop welding may be done by the shielded metal arc process, the submerged arc welding process, the gas metal arc welding process, or the flux core metal arc welding process.

Plates and component members of the tank should be assembled and welded following erection methods that minimize the amount of distortion due to weld shrinkage. The weld metal should meet or exceed the minimum requirements for visual inspection in AWWA D100 Sec. 11.4 and as set forth in the radiographic acceptance standards of AWWA D100 Sec. 11.6.10 (or other similar appropriate standards). Surfaces to be welded must be free from loose scale, slag, heavy rust, grease, paint, and other foreign material.

Tank Weld Testing

Weld testing is necessary as a quality assurance or quality control measure and is used to identify any incomplete (bad) welds as well as any welding process problems as soon as possible. The appropriate level of weld inspection will vary depending on the parameters of the tank (e.g., seismic use group, importance factor). Welding should be inspected and tested by radiographic testing consistent with the requirements of Sec. 11 of AWWA

D100, or other similar, appropriate standard. All welds should be tested in accordance with the most stringent testing regime allowed under AWWA D100, that is, all tests and inspections defined in AWWA D100 as being at the discretion of the "purchaser" should be carried out, and the resulting reports and other documentation should be furnished to the facility representative. The documentation includes, but is not limited to, mill inspection, shop inspection and field inspection as defined in AWWA D100.

Painting

All tank painting should be in accordance with Standards such as AWWA D102, the Steel Structures Painting Council Specification, SSPC-PA1, approved paint manufacturer's specifications, the thermal storage tank specification, and the separate project painting specifications.

Surface Preparation (roughness and cleanliness) is critical for the painting and coating system to be effective, since any weld metal splatter, sharp edges, oil or grease, and rust can prevent complete coating or adhesion to the metal surface. Therefore, the thermal storage tank specification should specify surface cleanliness minimum requirements such as, for example, prior to coating, all steel surfaces shall be prepared by abrasive blast cleaning; abrasive blast cleaning shall be carried out in accordance with the requirements of SSPC-SP10, and the standard of preparation shall be to SSPC-SP10 (or other appropriate standard). Further, the surface to be coated should meet the specified standard (i.e., be present on the surface) at the time of application of the first coat of paint. If the standard of preparation is not present, then the entire surface of the component should be re-cleaned in accordance with the thermal storage specification. To reduce surface roughness, the abrasive used in blast cleaning should achieve a surface roughness as recommended and stated on the paint manufacturer's data sheet.

The thermal storage tank specification should state that the contractor shall conform to all environmental regulations while tank blasting. Note, that pre-construction primers may be utilized in the fabrication process to preserve the blast profile and cleanliness.

All paint materials should be stored and used in strict compliance with the paint manufacturers recommendations and instructions. These instructions should be submitted on product data sheets and should form part of the thermal storage tank specification. As a minimum, the paint product data sheet should contain the following information:

- Required surface roughness profile
- Environmental limits for application, steel surface temperature, relative humidity, and air temperature
- Environmental limits during curing; steel surface temperature, relative humidity, and air temperature
- Minimum and maximum over-coating times over a range of temperatures
- Minimum times for handling and development of full cure over a range of temperatures
- Method(s) of application and number of coats required to achieve the specified thickness

No paint should be applied when the temperature of the surface to be painted is below the minimum temperature specified by the paint manufacturer or less than 5°

above the dew point temperature. Paint should not be applied to wet or damp surfaces or when the relative humidity exceeds 85 percent unless allowed by manufacturer's data sheets. The contractor must follow the paint manufacturer's recommendations for the specific paint system used.

All outside surfaces that have been welded, abraded, or otherwise damaged, should be blast cleaned per SSPC-SP 6. Inside damaged or uncoated surfaces should be blast cleaned per SSPC-SP 10, as noted. Primer should be applied in the field in accordance with the paint system requirements.

All areas blasted in the field must be coated the same day before any rusting occurs. The tank interior should be left in a broom-clean condition after completion of painting. All damage to applied coatings should be made good using the same method of preparation and coating materials as per the original application. Finally, prior to start of painting, the contractor should provide full containment for all sandblasting and painting operations.

Tank Coating and Inspection   Testing and inspection should be carried out in accordance with Sec. 5 of AWWA Standard D102, or other appropriate similar standard. The wet film thickness measurements must not be used for the calculation of dry film thickness. The contractor should monitor, record, and report the environmental conditions during coating application. Measurement of air and steel temperature, relative humidity and dew point calculation should be carried out three times per working shift as a minimum.

Hydrostatic Test

The thermal storage tank specification should require the tank contractor, in accordance with AWWA D100 (or other appropriate similar standard) to hydrostatically test the tank after the completion and acceptance of the tank coating after the tank coating has fully cured. The tank contractor should clean the tank prior to hydrostatic testing. External insulation should not be applied until the hydrostatic test is complete, and the thermal storage tank shows no leakage, as defined below. Upon completion of an acceptable hydrostatic test, the test water should be retained for process use, if possible.

The thermal storage tank should show no leakage for 48 hours when filled with water to the overflow. Should the tank show leakage, it should be drained and the leak repaired at no additional cost to the facility owner. Surfaces should be protected during the repair, and if damaged, should be restored to "as-new" condition. In the event of a repair, subsequent refilling, chemical addition, and retesting should be entirely at the contractor's expense.

As soon as it is confirmed that the thermal storage tank does not leak, the thermal storage tank contractor should provide the means to introduce the hydronic system (water treatment) chemicals. The project contractor should pay the cost of all chemicals used to treat the thermal storage volume (unless the facility owner prefers to pay for chemical costs directly).

Insulation

Before insulation application begins, all welding, leak testing, and painting on the vessel must be complete. If the contractor, at his risk, elects to insulate the tank prior to leak testing, any costs associated with the removal and reinstallation of insulation required to locate and repair leaks should be borne solely by the contractor.

The insulating boards should be installed as single or multiple layers. The edges of the boards should be tightly butted against each other. The joints of adjacent layers should be offset a minimum of 6 in. both vertically and horizontally, during installation.

Horizontal seams between jacket sheets (if required) on the tank shell and the seams at the top and bottom of the shell should be sealed vapor tight with a non-hardening, permanently flexible sealant or other approved material. Any lapped seams should be installed in the down weather position to allow water runoff.

Circumferential joints in the roof system (if required) should consider expansion joints and a down-weather lap position. The roof panels can be placed in a radial pattern or a "chevron" pattern as required to best address the tank diameter.

A reinforced metal corner flashing should be installed at the transition of the roof to shell jackets. This flashing section should be a rolled section (as opposed to segmental strips). Care should be taken to minimize penetrations through this corner flashing. Corner flashing should be adequately sealed and insulated.

Penetrations through the metal jacket should be sealed with a non-hardening flexible butyl sealant (or approved equivalent), and penetrations on the roof should be sealed using a "pitch pot" detail.

The perimeter area where the tank shell meets the foundation should be adequately sealed and insulated to minimize condensation. This should be accomplished by securing insulation and flashing to the tank shell jacket and to a rolled angle located outboard of the tank bottom secured to the concrete foundation with concrete anchors.

Seams between jacket sheets should be sealed vapor tight. Any lapped seams should be installed to allow water runoff.

Thermal and Pressure Drop Performance Tests

Part 3 of the thermal storage specification also identifies the tank testing requirements. Note that the thermal storage system (not just the tank, but the whole integrated system) testing requirements are specified in other separate project specifications, including test and balance, and controls specifications.

In the presence of facility representative, the thermal storage tank performance tests should be conducted to verify the thermal storage tank's ability to meet the performance requirements outlined in the thermal storage tank specification. The test should take place with the system control in automatic where possible.

The thermal storage tank manufacturer should provide a qualified engineer to witness the test, review the test data, and submit a written report to facility representative.

Performance testing may be conducted using facility installed, calibrated, NIST traceable instrumentation, including flow meters, tank inlet and outlet temperature sensors, and pressure sensors.

Prior to conducting any performance test, the contractor should submit to the facility representative a complete and detailed testing plan for review. All comments on the submittal should be fully resolved prior to scheduling the tests. The submittal should call out the proposed testing instrumentation, should indicate planned methodology of performing the tests, and should list the specified criteria for determining whether the tested performance has been satisfactorily achieved (e.g., the minimum required ton-hour capacity, heat gain, and pressure drop).

The thermal storage system will have met the specified performance requirements if satisfactory performance is demonstrated for two complete charge/discharge cycles (each test is run twice). The testing should be completed within 6 months after tank construction is complete, so as not to hold the tank contractor's retention too long.

Performance test preparation includes having the contractor notify the facility representative at least 2 weeks prior to the test date. For cooling thermal storage systems, prior to the performance test, the owner or facility representative should chill at least the lowest 8 ft of the water in the storage tank to the design supply temperature and hold that temperature for 3 days to confirm that the soil and foundation have reached a steady state thermal condition.

The facility representative should also confirm that inlet/outlet flow meter, inlet/outlet temperature instrumentation, inlet/outlet pressure drop instrumentation, and data collection devices are operational and calibrated.

Test Procedure No. 1: Regeneration Test (Tank Charging)   The initial test of the measured thermal performance should be based upon regenerating the thermal storage system. Prior to the start of the performance test, the owner or facility representative should bring the thermal storage tank to a steady state, fully discharged condition, with only residual cold water, if any, in the case of chilled water, at the bottom of the tank as determined by the thermal instrumentation.

The owner or facility representative should perform the regeneration performance test by withdrawing warm water, at the specified design temperature, from the top of the thermal storage tank at various rates, up to the maximum design flow rate. If the warm water temperature is higher than the specified design temperature, assume the specified design temperature for calculating the storage capacity (the contractor should not take credit for water warmer temperature than specified in the thermal storage tank). Return water at the specified design temperature to the bottom of the thermal storage tank at an equal flow rate to that withdrawn from the top of the tank. The inlet/outlet flow rates and temperatures should be recorded at 1- minute intervals. Record the temperatures along the vertical dimension of the tank at 15-minute intervals to verify the thermocline performance and thickness. Calculate the thermal load delivered into storage during these intervals from the flow and temperature recordings. The total stored thermal energy should be the capacity sum of all periods during the test. Chart the recorded performance factors.

In the case of CHW thermal storage, the thermal storage tank should be considered as fully charged when the top surface of the thermocline reaches the warm water nozzle, which should be determined by measuring the warm water temperature at the nozzle, and by tracking the thermocline location with the tank thermocouples. Note, during actual operations, the entire thermocline can be withdrawn leaving the thermal storage tank completely filled with cold water.

The thermal storage tank will have met the specified storage performance requirements for regeneration if the total thermal content (e.g., in ton-hours) in the thermal storage tank is equal to or greater than the specified design value (within the stated accuracy of the flow and temperature instrumentation). The regeneration test should be performed twice for comparison purposes.

Test Procedure No. 2: Depletion Test (Tank Discharging)   The depletion test should be based upon discharging the thermal storage tank. Prior to the start of the discharge performance test, the owner or facility representative should bring the thermal storage tank to a steady state fully charged condition, as determined by the thermal storage instrumentation. The discharge performance test should be conducted at such time as sufficient thermal load on the hydronic system is available to permit the complete discharge of the system at the maximum specified design flow rate, unless alternative arrangements have been made.

The owner or facility representative should perform the discharge performance test by withdrawing chilled water from the bottom of the tank at various rates, up to the maximum design rate. Warm water should be returned to the top of the tank at the specified design temperature and at an equal flow rate to that withdrawn. If the warm water temperature is higher than the specified design temperature, assume the specified design temperature for calculating the storage capacity (the contractor should not take credit for warmer return water temperature than specified). Record the inlet/outlet flow rates and temperatures at 1-minute intervals. Record the temperatures along the vertical dimension of the tank at 15-minute intervals to verify the thermocline performance and thickness. Calculate the thermal load extracted from the thermal storage system during these intervals from the flow and temperature recordings. Chart the recorded performance factors.

The thermal storage tank should be considered as fully discharged when the bottom surface of the thermocline reaches the cold water nozzle, which should be determined by measuring the chilled water temperature at the nozzle and by tracking the thermocline location with the thermocouples. Moreover, similar to the charging mode, in actual operation, the entire thermocline can and likely should be withdrawn.

The thermal storage tank will have met the specified performance requirements for discharge if the total thermal capacity (e.g., ton-hours) withdrawn from the tank is equal to or greater than the specified design value within the accuracy of the flow and temperature instrumentation. The depletion test should be performed twice.

Test Procedure No. 3: Pressure Drop Test   The thermal storage tank specification should require the contractor to measure the pressure drop from the tank cold water inlet/outlet flange to the warm water inlet/outlet flange at the maximum specified discharge rate and charge rate. The measured pressure drop should be less than the specified maximum allowable pressure drop. If the pressure drop across the tank is greater than the specified drop, the tank contractor must modify the tank internals to provide the specified drop without degrading the thermal performance of the thermal storage tank.

Test Procedure No. 4: Heat Gain Verification Test, Conducted with No Flow in the Tank   Maintain the thermal storage tank in a fully charged, stagnant condition for a number of hours, while recording the ambient air temperature at 15-minute intervals. The heat gain (for cooling systems) in storage during the holding test will be calculated from the tank temperature recordings at the start and end of the holding period. The total heat gain rate will be calculated using the recorded ambient air temperatures and then adjusted to predict heat gain at the design ambient air temperature.

Operation and Maintenance Data

The contractor should provide the required number of copies of the Operation and Maintenance (O&M) manual to the facility representative, and the manual should be delivered 30 days prior to substantial completion of field construction.

As a minimum, the O&M manual should include the following:

- Thermodynamic description of the charge and discharge cycles
- Safety precautions and instructions

- Detailed performance test procedures
- Recommended spare parts list
- Recommended maintenance practices and schedules for the tank foundation, structure, appurtenances, insulation, internal distribution system, and special systems such as instrumentation
- Local contact for the thermal storage tank contractor, and for other subcontractors and suppliers of major components

Terms of Final Acceptance

The final acceptance of the thermal storage tank by the facility representative should be made when the performance requirements of the tank have been demonstrated to meet the requirements of the thermal storage specification.

The thermal storage specification should state that the tank contractor shall modify the system at no increase in contract sum to correct any deficiencies in performance. If there is a deficiency in the thermal performance, one option may be to allow the contractor to choose his or her option to provide compensation as previously described in lieu of correcting the deficiency. The tank contractor should reimburse the facility owner for reasonable direct costs for professional engineering services incurred to assess and correct any deficiencies, and for any direct costs incurred by the facility for re-testing required to validate corrective work.

Above- or Belowground Concrete Thermal Storage Tank Specifications

The following outlines and discusses concrete thermal storage tank specifications that are different from or expanded upon that described previously for an aboveground steel thermal storage tank. Specification information needed for a concrete thermal storage tank specification that is the same as or similar to a steel thermal storage tank is not repeated here again. The main differences between the aboveground steel tank and the concrete thermal storage tank are the tank material and method of construction; foundation requirements; insulation fastening methods, and instrumentation mounting requirements.

Different construction methods are used to build different types of concrete thermal storage tanks including the following: conventional, poured-in-place, reinforced concrete; precast, prestressed wire wound concrete with steel diaphragm; prestressed strand-wound cast-in-place concrete, and tendon prestressed concrete tanks, for example. However, it should be noted that, because the chilled water stored in the tank is chemically treated, it is imperative that the tank construction be watertight. The word "wire" refers to either steel wire or galvanized seven-wire strand that is wrapped circumferentially around the thermal storage tank. "Prestressed" refers to the introduction of stresses to a structural member that counteract the stresses resulting from self-weight and superimposed loadings (concrete is very strong in compression but weak in tension). Pre-tensioning is the method of prestressing structural components when concrete production occurs at a precast concrete facility. Post-tensioning is the method of prestressing when concrete members are cast in place at the project site.

Different tank construction methods are governed by different tank construction standards. AWWA D110, Wire- and Strand-wound, Circular, Prestressed Concrete Water

Tanks, governs wire-wrapped prestressed concrete tanks. Per the AWWA Web site, the intent of Standard D110 is to describe current recommended practice for the design, construction, inspection, and maintenance of wire- and strand-wound, circular, prestressed concrete water-containing structures with the following four different types of core walls:

- Type I—cast-in-place concrete with vertical prestressed reinforcement
- Type II—shotcrete with a steel diaphragm
- Type III—precast concrete with a steel diaphragm
- Type IV—cast-in-place concrete with a steel diaphragm (however, currently, there are no thermal storage tank builders using this methodology)

AWWA Standard D110 does not describe bonded or unbonded horizontal tendons for prestressed reinforcement of the tank wall, floor, or roof, and applies to containment structures for use with potable water or raw water of normal temperature and pH commonly found in drinking water supplies.

AWWA D115, Standard for Tendon-Prestressed Concrete Water Tanks, is the standard used for tendon prestressed concrete tanks and describes current and recommended practice for the design, construction, and field observations of concrete tanks using tendons for prestressing, and applies to containment structures for use with potable water, raw water, or wastewater. The American Petroleum Institute (API) also writes tank design and construction standards, and some thermal storage tanks have been designed to their standards (API 650, e.g.).

Part 1—Execution

Part 1 of the concrete thermal storage tank specification is very similar to the steel thermal storage tank specification, as the Part 1 paragraphs describe the general thermal storage system requirements, except, of course, that the design and construction references are regarding concrete not steel (e.g., no welding, no grounding lugs, no lightening protection). The description of the concrete thermal storage tank may, for example, be as follows: "The thermal storage tank shall consist of a cast-in-place prestressed concrete wall with vertical post-tensioning or precast concrete wall panels with integral steel diaphragm (or tank could be a cast-in-place concrete flat roof with interior columns as needed to support the design roof loadings or precast dome panels for a free roof span), an internal flow diffuser system, and thermal insulation (if needed to supplement the inherent insulating value of the concrete and surrounding soil)."

The following are typical codes and standards for AWWA-constructed water tanks:

- ACI 301 Specifications for Structural Concrete for Buildings
- ACI 305 Hot Weather Concreting
- ACI 306 Cold Weather Concreting
- ACI 309R Guide for Consolidation of Concrete
- ACI 318 Building Code Requirements for Reinforced Concrete and Commentary
- ACI 350 Code Requirements for Environmental Engineering Concrete Structures and Commentary

- ACI 350.3 Seismic Design of Liquid Containing Concrete Structures and Commentary
- ACI 372R Design and Construction of Circular Wire- and Strand Wrapped Prestressed Concrete Structures
- ACI 373R Design and Construction of Circular Prestressed Concrete Structures with Circumferential Tendons
- ACI 506R Guide to Shotcrete
- ASHRAE HVAC HVAC Systems and Equipment Handbook, Chap. 50, Thermal Storage
- ASHRAE Design Guide for Cool Thermal Storage
- ASTM A185 Specification for Steel Welded Wire, Fabric, Plain for Concrete
- ASTM A416 Standard Specification for Steel Strand, Uncoated Seven—Wire for Prestressed Concrete
- ASTM A475 Standard Specification for Zinc-Coated Steel Wire Strand
- ASTM A615 or A615M Specification for Deformed and Plain Billet-Steel Bars for Concrete Reinforcement
- ASTM A821 Specification for Steel Wire, Hard Drawn for Prestressing Concrete Tanks
- ASTM A1008/A1008M Specification for Steel, Sheet, Cold-Rolled, Carbon, Structural, High-Strength Low-Alloy and High-Strength Low-Alloy With Improved Formability
- ASTM C31 Standard Practice for Making and Curing Concrete Test Specimens in the Field
- ASTM C33 Standard Specification for Concrete Aggregates
- ASTM C39 Standard Test Method for Compressive Strength of Cylindrical Concrete Specimens
- ASTM C618, Type F Standard Specification for Coal Fly Ash and Raw or Calcined Natural Pozzolan for Use in Concrete
- ASTM C920 Specification for Elastomeric Joint Sealants
- ASTM D1056 Standard Specification for Flexible Cellular Materials—Sponge or Expanded Rubber
- ASTM D1556 Standard Test Method for Density and Unit Weight of Soil in Place by the Sand-Cone Method
- ASTM D1557 Standard Test Methods for Laboratory Compaction Characteristics of Soil Using Modified Effort (56,000 Ft-lbf/ft3) 2700 KN-M/M3)
- ASTM D2000 Classification System for Rubber Products in Automotive Applications
- ASCE Standard 7-05 Minimum Design Loads for Buildings and Other Structures
- AWWA C652 Standard for Disinfection of Water-Storage Facilities

- AWWA D110 Wire and Strand Wound, Circular, Prestressed Concrete Water Tanks (list type no. 1, 2, or 3)
- AWWA D115 Standard for Tendon-Prestressed Concrete Water Tanks
- U.S. Army Corps of Engineers Specification CRD-C-572, Specification for PVC Waterstop

Specific design criteria that should be specified for concrete tanks include that horizontal prestressing should be continuous, and discontinuous prestressing tendons or strands should not be allowed. Further backfill pressure (earth loads) should be determined by methods of soil mechanics, and backfill pressure should not be used to reduce the amount of required prestressing. The tank foundation must be proportioned so that soil pressure is less than the allowable soil bearing capacity.

The thermal storage tank should be designed and constructed in accordance with AWWA D110 (or similar appropriate standard) seismic zone for the particular tank location, including an accommodation for the seismic "sloshing wave" through the use of either appropriate freeboard between the normal operating water surface and roof or a roof capable of resisting the uplift of such a seismic wave. The tank may also be designed in accordance with the International Building Code (IBC) Site Class for the specific location and as specified by the geotechnical engineer in the site-specific seismic guidelines. Note that the minimum requirements for tendon tanks are different from the minimum requirements for wire- and strand-wrapped tanks.

As determined by an SE in practice, minimum cast-in-place wall thickness should be 12 in. for the full height of the wall. For penetrations greater than 24 in. in height, the wall should be designed to provide adequate local reinforcement for non-uniform prestressing distribution on the wall.

Further, the floor slab should be designed as a slab on grade not less than 6-in. thick with two mats of reinforcing and should be placed monolithically. No construction joints should be allowed unless otherwise approved by the engineer. Wall footings may be above or below floor grade, but should be placed monolithically with the floor.

A flat, column-supported roof should have a minimum roof thickness of 12 in. (or other thickness as determined by the SE). The flat roof shall be allowed to radially expand and contract, but should be restrained in the tangential direction to the tank wall. Columns or interior supports should be a minimum 18 in. in diameter (per SE), spirally reinforced, as required to support the roof loads, while accommodating the design of the internal flow diffusers. Both the flat column-supported roof and columns should be designed in accordance with ACI 350-06 requirements (or other similar appropriate standard). Also, a free span dome roof can also be constructed of 4-in. thick precast concrete panels, or 4-in. thick cast-in-place concrete. A free span dome roof is typically more prevalent for partially buried or aboveground tanks.

In comparison to the steel tank submittals, concrete tanks submittals should also include the following:

- Design proportions for all concrete and shotcrete and concrete strengths of trial mixes
- Admixtures to be used in the concrete and their purpose
- Reinforcing steel shop drawings showing fabrication and placement

Part 2—Material

Part 2 describes or specifies tank materials and appurtenances. The following provides general basic specifications for a concrete tank.

Concrete

The following provides a basic example of a concrete specification:

- Concrete shall conform to ACI 301.
- Cement shall be Portland cement Type II or Type V.
- Admixtures, other than air-entraining and water-reducing admixtures, shall not be permitted unless approved by the engineer; the maximum water-cement ratio for all concrete shall not exceed 0.42.
- Concrete for tank wall and roof construction shall have a minimum compressive strength of 4000 psi at 28 days. All wall and roof concrete shall be air-entrained.
- Concrete for the tank floor, footings, pipe encasement, and all other work shall have a minimum compressive strength of 4000 psi at 28 days and shall not be air-entrained.
- The course and fine aggregate shall meet the requirements of ASTM C33; course aggregate shall be No. 467 with 100 percent passing the 1- to 1/2-in. sieve; superplasticizer, water-reducing, and shrinkage compensating admixtures shall be incorporated into the floor concrete.
- Proportioning for concrete shall be in accordance with ACI 301.
- Concrete or shotcrete in contact with prestressing steel shall have a maximum water soluble chloride ion concentration in the concrete or shotcrete of 0.06 percent by weight of cement.

Shotcrete

The following provides a basic example of a shotcrete specification:

- Shotcrete shall conform to ACI Standard 506, except as modified herein.
- Wet mix process shall be employed for shotcreting.
- Shotcrete used for covering prestressed wire shall consist of not more than three parts sand to one part Portland cement by weight. Additional coats of shotcrete shall consist of not more than four parts sand to one part Portland cement by weight. Shotcrete shall have a minimum strength of 4500 psi at 28 days.
- All shotcrete shall be fibrous reinforced; such material shall consist of 100 percent virgin polypropylene fibrillated fibers in accordance with ACTM C-116.

Mortar Fill and Nonshrink Grout

Mortar fill and nonshrink grout should have a minimum compressive strength of 4000 psi at 28 days.

Reinforcing Steel

The following provides a basic example of a reinforcing steel specification:

- Reinforcing steel shall be new billet steel Grade 60, as shown in the drawings, meeting the requirements of ASTM A615; welded wire fabric shall conform to ASTM A185.
- Reinforcing steel shall be accurately fabricated and shall be free from loose rust, scale, and contaminants that reduce bond.
- Reinforcing steel shall be accurately positioned on supports, spacers, hangers, or other reinforcements and shall be secured in place with wire ties or suitable clips; rebar chair supports may be either steel with plastic tips or plastic.
- The tank designer shall use base restraint cables to resist earthquake loads; seismic base restraint cables shall be hot-dipped galvanized seven-wire strand and shall be manufactured in accordance with ASTM A416 prior to galvanizing, and ASTM A475 after galvanizing.

Vertical Post-Tensioning (for AWWA D110 Type 1 Concrete Tanks)

The following provides a basic example of a vertical post-tensioning specification used in AWWA D110 Type I concrete tanks:

- Deformations of the thread bars shall form a screw thread suitable for mechanically coupling lengths of thread bar and for positive attachment of anchor assemblies.
- Deformations shall conform to ASTM A722, Type II requirements and shall be uniform such that any length of bar may be cut at any point, and the internal threads of a coupling designated for that size of bar can be freely screwed on the bar; the bars and their deformations shall be hot rolled.
- Tensile and Physical Properties shall be manufactured in accordance with ASTM A 722, Type II.

Steel Diaphragm:

- The steel diaphragm shall conform to ASTM A1008, shall be a minimum thickness of 0.017 in., and shall be vertically ribbed with reentrant angles; the back of the channels shall be wider than the front, thus providing a mechanical keyway anchorage with the concrete and shotcrete encasement.
- The steel diaphragm shall extend within 1 in. of the full height of the wall panel with no horizontal joints, and vertical joints within a wall panel shall be roll seamed or otherwise fastened in a fashion which results in a firm mechanical lock; joints between wall panels that are not roll seamed shall be edge sealed with polysulfide or polyurethane sealant.
- No punctures will be permitted in the diaphragm except those required for pipe sleeves, temporary construction openings, or special appurtenances; details of such openings, as are necessary, shall be approved by the engineer,

and all such openings shall be completely edge sealed with polysulfide or polyurethane sealant.

- Diaphragm steel may be considered as contributing to the vertical reinforcement of the wall.

Galvanized Prestressing Steel (for AWWA D110 Type 1 Tanks)

Steel for prestressing can be cold drawn, high carbon wire meeting the requirements of ASTM A821, hot-dipped galvanized having a minimum ultimate tensile strength of 210,000 psi, or hot-dipped galvanized seven-wire strand manufactured in accordance with ASTM A-416 prior to galvanizing. Each wire of the strand should be individually hot-dipped galvanized before being stranded.

Splices for horizontal prestressed reinforcement should be ferrous material compatible with the reinforcement and shall develop the full strength of the wire. Wire splice and anchorage accessories should not nick or otherwise damage the prestressing.

Post-Tensioning Systems

1. All post-tensioning components and systems shall be furnished by a PTI-certified plant.
2. Strand shall be low relaxation, seven-wire strand, having a guaranteed minimum ultimate tensile strength of 270,000 psi, per ASTM Designation A 416, low relaxation.
3. Bonded Tendons:
   a. Anchorages must develop at least 95 percent of the guaranteed ultimate tensile strength (GUTS) of the prestressing steel, tested in an unbonded state, without exceeding anticipated wedge seating. Anchorages shall include details that completely encapsulate the post-tensioning steel with mechanical connections; friction-type connections will not be allowed.
   b. Duct for bonded tendons shall be manufactured from high-density polypropylene (HDPP) materials that completely encapsulate the post-tensioning steel and include mechanical type couplers.
   c. Grout for injection grouting of bonded tendons shall be Prepackaged Class C per PTI Specifications.
4. Unbonded Tendons:
   a. Anchorages must develop at least 95 percent of the guaranteed ultimate tensile strength (GUTS) of the prestressing steel without exceeding anticipated wedge seating. Anchorages shall include details that completely encapsulate the post-tensioning steel with mechanical connections; friction-type connections will not be allowed.
   b. The sheathing for unbonded tendons shall manufactured from high-density polyethylene (HDPE) and shall be produced by a seamless extrusion process over corrosion preventative grease that leaves no air pockets and has at least a 50-mil thickness with no negative tolerance.
   c. The sheathing at the anchorage shall be encased by a watertight plastic trumpet filled with corrosion preventative grease. Exposure of bare strand shall not be permitted.

d. Extruded strand bulk coils for unbonded tendons shall be packaged immediately upon completion of extrusion to prevent handling damage. Tendons for floor and roof slabs shall be fabricated at the job site from bulk coils and not plant fabricated.

e. Sheathing repairs, if required, shall be in accordance with PTI recommendations.

Tank Appurtenances

Concrete tank appurtenances are similar to steel tank appurtenances except that temperature sensors are mounted from the roof of the concrete tank (versus along the wall of a steel tank).

Typical appurtenances in concrete tanks are as follows:

1. Wall insulation

2. Diffuser piping

3. Overflow weir and piping sized based on the maximum inlet flow

4. *Roof Hatch*: A 48-in. minimum square aluminum hatch with lockable, hinged cover and curb frame. The hatch shall have a lift handle, padlock tab, padlock, and a cover hold open mechanism, and all hardware shall be aluminum or stainless steel.

5. *Roof Ventilator*: Fiberglass or Aluminum, with fiberglass insect 20 × 20 screen, minimum diameter 2 ft 0 in.

6. *Interior Ladder*: A fiberglass or aluminum ladder shall extend from the hatch to the floor. The ladder shall have a fall prevention device attached consisting of a sliding, locking mechanism and safety belt and complying with applicable OSHA standards.

7. *Exterior Ladder*: An aluminum ladder shall extend from 8 ft above the final grade to the tank roof. The ladder shall have an OSHA-approved fall prevention device (if required) consisting of a sliding, locking mechanism and safety belt.

8. *Access Manway*: A circular 25-in. diameter Type 304 stainless steel wall manway with a hinged cover. A Type 304 stainless steel (S.S.) grab bar and an aluminum ladder shall be installed at the manway location.

9. *Temperature Sensor Sleeve*: 6-in. diameter S.S. or D.I. pipe with flange end extending above and below the tank roof top and bottom surfaces a minimum of 4 in.

10. *Temperature Sensor Protector*: 6-in. diameter perforated PVC pipe installed vertically inside the TES tank attached to the tank wall (minimum 3 ft from wall) to receive the individual sensors (Please note that SS wall sleeves can be provided for mounting individual temperature probes; however, this is typically more expensive than the roof mounted arrangement.)

a. *Level Sensor Sleeve*: 6-in. diameter S.S. or D.I. pipe with flange end extending above and below the tank roof top and bottom surfaces a minimum of 4 in.

Part 3—Construction

Where any concrete foundation or concrete slab on grade will be used, and especially for belowground thermal storage tanks, excavation and backfill must be carefully specified including using separate project earthwork specifications. As the installation means and methods for concrete are critical to the performance of the final concrete product, Part 3 of the concrete thermal storage tank specification describes the proper placement and curing of concrete and shotcrete. Similar appurtenance items specified for steel thermal storage tanks are also specified.

Part 3 should specify for the contractor to first clear and grub the proposed tank site by removing all trees, shrubs, brush, stumps, roots, and other unsuitable material to a minimum distance of 10 ft outside the edge of the tank foundation, plus any additional areas necessary for the tank construction. The limits of clearing should be as shown on the drawings or as approved by the engineer. All trees and vegetation should be disposed of off-site, and all topsoil shall be stripped from the proposed construction work area and stockpiled on-site. AWWA D115 post-tensioning construction requirements have been included in the Appendices.

Excavation and Backfill

The following provides a basic example of excavation and backfill specifications:

A. The contractor should excavate to such depths and widths to provide adequate room for tank construction. A minimum working area of 10 ft beyond the circumference of the tank foundation at an elevation 6 in. below the top of the tank foundation shall be provided. Excavated material may be used as suitable backfill material and stockpiled on site as required.

B. The excavation must be dewatered as required during construction. The dewatering method used shall prevent disturbance of the tank foundation soils. A permanent dewatering system will be provided by the contractor, as appropriate to suit the soil conditions described in the available geotechnical report, for use during initial construction, normal operation, and potential subsequent out-of-service events when the tank could be emptied.

C. The contractor shall excavate rock, if encountered, to the lines and grades indicated on the drawings, or as directed by the engineer. Rock excavation should be measured separately and paid for by the unit price item for rock excavation indicated in the bid. The pay limit for rock in the area of the tank should be carried out to 10 ft beyond the circumference of the tank foundation and at an elevation of 1 ft below the tank foundation.

D. In the event the subgrade material is disturbed or over-excavated by the contractor during excavation, it must be removed and replaced with compacted select fill, at the contractor's expense.

E. If, in the opinion of the engineer, the subgrade is unsuitable for the foundation, the engineer should direct that it be removed by the contractor and replaced with compacted select fill. Unsuitable material and compacted select fill shall be measured separately and paid for by the unit price indicated in the bid.

F. After excavation is complete, the bottom of the excavation must be proof rolled and leveled as directed by the engineer before the compacted select fill is placed. The engineer should inspect the subgrade for conformance with the original geotechnical report and its suitability for the tank foundation. Before any select fill is to be placed against rock surfaces, the rock must be relatively free of all vegetation, dirt, clay, boulders, scale, excessively cracked rock, loose fragments, ice, snow, and other objectionable substances. All free water left on the surface of the rock must be removed.

G. A leveling base material consisting of a minimum 6-in. thick layer of compacted select fill should be placed beneath the entire tank foundation. Select fill should be placed in layers not exceeding 12 in. and compacted to a minimum density equal to 95 percent of the maximum laboratory density in accordance with ASTM D1557 or other appropriate standard. A nonwoven geotextile fabric shall be placed between the subgrade and leveling base material as shown on the drawings or directed by the tank builder. Select fill shall consist of a clean, well-graded angular or subangular material having not more than 8 percent by weight passing the No. 200 sieve. The maximum size stone shall be 1½ in. If directed by the tank builder, a uniformly graded ¾-in. crushed stone must be used as the leveling base material. The crushed stone shall be ¾-in. sieve size with 100 percent passing the 1 in. If uniformly graded crushed stone is used for the leveling base material, compaction performance criteria must be used to gauge the degree of compaction. Crushed stone should be placed in layers not exceeding 9 in. and compacted with at least two passes in each direction with vibratory roller compaction equipment. Field testing for density achieved shall be in accordance with ASTM D1556 or D2922.

H. The surface elevation of the compacted select fill should be fine graded to a tolerance of plus +0 to −½ in. over the entire foundation areas. Fine grading tolerances for floor pipe encasements shall be +0 to −6 in.

I. The tank should be backfilled and rough graded to the contours shown on the drawings. Unless other material is specified by the engineer, materials used for backfilling shall be suitable on site material.

J. Frozen material must not be used for backfill nor shall fill material be placed on snow, ice, or frozen material. Rock or concrete spoils (greater than 6 in.) must not be used in backfill within 2 ft of the tank wall.

Tank Floor Construction

The following provides a basic example of a concrete tank floor specification:

A. The floor and wall footings must be constructed to the dimensions shown on the approved shop drawings.

B. Prior to placement of the floor, a 6 mil polyethylene moisture barrier should be placed over the subbase. Joints in the polyethylene must be overlapped a minimum of 6 in.

C. Prior to placement of the floor, all piping that penetrates through the floor, must be set and encased in concrete.

D. The vertical waterstop must be placed and supported so that the bottom of the center bulb is at the elevation of the top of the footing. The waterstop must be supported without puncturing any portion of the waterstop, unless it is manufactured with holes for tying. The waterstop should be spliced using a thermostatically controlled sealing iron and each splice should be successfully spark tested prior to encasement in concrete.

E. The floor shall have a minimum thickness of 6 in. (note again actual floor thickness as determined by SE) and be poured monolithically. There shall be no construction joints in the floor or between the floor and footing. Floors over 30,000 sq. ft in surface area may, at the option of the contractor, have one or more construction joints. Such construction joints must be approved by the engineer prior to placement.

F. The floor shall be cured by applying one coat of curing compound and water for a period of 7 days.

Precast Panel Construction and Erection

The precast wall should be constructed with a continuous waterproof steel diaphragm embedded in the exterior of the precast panel. Horizontal joints in the diaphragm should not be allowed.

No holes for form ties, nails, or other punctures should be permitted in the wall.

Temporary wall openings may be provided for access and removal of construction materials from the tank interior subject to the approval of the engineer.

Wall and dome panel beds should be located around the periphery of the tank as required. The beds must be constructed to provide finished panels with the proper curvature of the tank.

Polyethylene sheeting should be placed between successive pours to provide a high moisture environment and a long slow cure for the concrete.

The erection crane and lifting equipment must be capable of lifting and placing precast panels to their proper location without causing damage to the panel.

The precast panels must be erected to the correct vertical and circumferential alignment. The edges of adjoining panels should not vary inwardly or outwardly by more than 3/8 in. and should be placed to the tank radius within + 3/8 in.

Joints between precast wall panels shall be bridged with a minimum 10-gauge steel plate, edge sealed with polysulfide, and filled with mortar. No through-wall ties should be permitted.

Minimum dome and wall panel thickness should be 4 in.

Concrete

The following provides a basic example of a concrete installation specification:

A. All concrete shall be conveyed, placed, finished, and cured as required by pertinent ACI standards.

B. Weather Limitations
   1. Unless specifically authorized in writing, concrete shall not be placed without special protection during cold weather when the ambient temperature is below 35°F and when the concrete is likely to be subjected to freezing temperatures before final set has occurred and the concrete strength has

reached 500 psi. Concrete shall be protected in accordance with ACI 306R. The temperature of the concrete shall be maintained in accordance with the requirements of ACI 301 and 306R. All methods and equipment for heating and for protecting concrete in place shall be subject to the approval of the engineer.

2. During hot weather, concreting shall be in accordance with the requirements of ACI 305R.

3. Placement of concrete during periods of low humidity (below 50 percent) shall be avoided when feasible and economically possible, particularly when large surface areas are to be finished. In any event, surfaces exposed to drying wind shall be covered with polyethylene sheets immediately after finishing, or flooded with water, or shall be water cured continuously from the time the concrete has taken initial set. Curing compounds may be used in conjunction with water curing, provided they are compatible with coatings that may later be applied, or they are degradable.

C. Finishes

The tank should be given the following finishes:

1. The floor slab shall be given a hard steel trowel finish.
2. The exterior of the cast-in-place roof shall be given a light broom finish.
3. Exterior shotcrete shall be given a fine texture gun finish.

D. Curing

1. Concrete shall be cured using water methods, sealing materials, or curing compounds. Curing compounds used on the exterior of the tank wall shall be removed by abrasive blasting the corewall prior to the application of the shotcrete. Curing compounds used within the tank shall be suitable for use with potable water.

E. Testing

1. For concrete placed in the cast-in-place wall, floor, or roof, two sets of five cylinders for each 50 cubic yards are placed on the same day. One cylinder shall be tested at 7 days, one at 14 days, two at 28 days, and one held as a spare.

2. Slump, air content, and temperature testing shall be performed on each truck where cylinders are taken.

3. All concrete testing shall be in accordance with ASTM C-31 and C-39, at the contractor's expense, and shall be conducted by an independent testing agency approved by the engineer.

Shotcreting

The following provides a basic example of a shotcrete construction specification:

A. Weather Limitations:

1. Shotcrete shall not be placed in freezing weather without provisions for protection of the shotcrete against freezing. Shotcrete placement can start without special protection when the temperature is 35°F and rising, and must be suspended when the temperature is 40°F and falling. The surface to

which the shotcrete is applied must be free from frost. Cold weather shotcreting shall be in accordance with ACI 301 and ACI 306R.
  2. Hot weather shotcreting shall be in accordance with the requirements of ACI 301 and ACI 305R.

B. Shotcrete Coating over Prestressing Wire:
  1. Each prestressed wire shall be individually encased in shotcrete. Shotcrete wire coat thickness shall be sufficient to provide a clear cover over the wire of at least 1/4 in.
  2. A finish coat of shotcrete shall be applied as soon as practical after the last application of wire coat. The total thickness of shotcrete shall not be less than 2 in. over the wire and shall be formed in 0.5 to 0.75 in. layers for the full height and circumference prior to beginning the subsequent layers.

C. Shotcrete Coating over Steel Diaphragm
  1. The steel diaphragm shall be covered with a layer of shotcrete at least 1/2 in. thick prior to prestressing.
  2. Total minimum coating over the steel diaphragm shall be 1 to 1/2 in. including diaphragm cover, wire cover, and finish cover coat.

D. Placement of Shotcrete:
  1. Shotcrete shall be applied with the nozzle held at a small upward angle not exceeding 5° and constantly moving during application in a smooth motion with the nozzle pointing in a radial direction toward the center of the tank. The nozzle distance from the prestressing shall be such that shotcrete does not build up or cover the front face of the wire until the spaces behind and between the prestressing elements are filled.
  2. Total shotcrete covercoat thickness for hand-applied shotcrete shall be controlled by shooting guide wires. Vertical wires shall be installed under tension and spaced no more than 2 ft apart to establish uniform and correct coating thickness. Wires of 18 or 20 gauge high tensile strength steel or a minimum 100 lb. monofilament line shall be used. Wires shall be removed after placement of the shotcrete covercoat and prior to finishing.
  3. All shotcrete shall be applied by ACI certified nozzlemen. Certifications shall be submitted to the engineer for review at bid time.

E. Curing:
  1. Shotcrete shall be cured using water curing methods, sealing materials, or curing compounds at the option of the contractor.

F. Testing:
  1. Testing of shotcrete shall be in accordance with ACI 506, except as specified herein. One test panel shall be made for each of the following operations: corewall, cove, wire cover, and covercoat. Test panels shall be made from the shotcrete as it is being placed, and shall, as nearly as possible, represent the material being applied. The method of making a test sample shall be as follows: A frame of wire fabric (1 ft2, 3 in. deep) shall be secured to a plywood panel and hung or placed in the location where shotcrete is being placed. This form shall be filled in layers simultaneously with the nearby application. After 24 hours, the fabric and plywood backup shall be removed and the sample slab placed in a safe location at the site.

2. The sample slab shall be moist cured in a manner identical with the regular surface application. The sample slab shall be sent to an approved testing laboratory and tested at the age of 7 days and 28 days. Nine 3-in. cubes shall be cut from the sample slab and subjected to compression tests in accordance with current ASTM Standards. Three cubes shall be tested at the age of 7 days, three at the age of 28 days, and three shall be retained as spares. Testing shall be by an independent testing laboratory, approved by the engineer and at the contractor's expense.

Prestressing

The following provides a basic example of a prestressing construction specification:

A. Prestressing wire should be placed on the wall with a machine capable of consistently producing a stress in the wire within a range of –7 percent to +7 percent of the stress required by the design. Stress shall be developed through mechanical tensioning of the prestressed wire. Stressing of the wire may not be accomplished by drawing the wire through a die or in any way deforming the wire. The temperature increase in the wire caused by the stressing method employed shall not exceed 50°F. No circumferential movement of the wire along the tank wall will be permitted during or after stressing the wire.

B. Each coil of prestressing wire shall be temporarily anchored at sufficient intervals to minimize the loss of prestress in case a wire breaks during wrapping.

C. Minimum clear space between prestressing wires is 5/16 in. or 1.5 wire diameters, whichever is greater. Any wires not meeting the spacing requirements shall be re-spaced. Prestressing shall be placed no closer than 2 in. from the top of the wall, edges of openings, or inserts, nor closer than 3 in. from the base of walls or floors where radial movement may occur.

D. The band of prestressing normally required over the height of an opening shall be displaced into circumferential bands immediately above and below the opening to maintain the required prestressing force. Bundling of wires shall be prohibited.

E. Ends of individual coils shall be joined by suitable steel splicing devices capable of developing the full strength of the wire.

F. The contractor must furnish a calibrated stress-recording device for both the vertical and circumferential prestressing operations, which can be recalibrated, to be used in determining wire stress levels on the wall during and after the prestressing process. All vertical and circumferential stress-recordings devices shall be calibrated at an approved independent lab not more than 14 days prior to being used in the field. Circumferential and vertical stress recordings shall be continuous and instantaneous as the wire is applied to the wall, and the stress recordings shall be plotted as quality assurance for the owner. If the applied stress in the circumferential or vertical steel is more than 7 percent under or over the required design stress, the stressing operation should be discontinued and satisfactory adjustment made to the stressing equipment before proceeding.

Liquid Tightness Test

Upon completion, the tank must be tested to determine liquid tightness. The tank should be filled with water to the maximum level. Water typically is furnished to the tank by the owner. The test should be in accordance with AWWA D-110-04, Sec. 5.13 (or other appropriate standard), except that the tank contractor is required to guarantee zero leakage (as required by Steel tanks constructed under AWWA D100).

Insulation

Depending upon the thickness of the concrete, whether the tank is aboveground, partially buried, or underground, the soil temperatures, and the outside air temperatures, insulation may be required in order to limit the maximum heat gain (in the case of a cooling system) to less than 2 percent of daily rated capacity. If insulation is used or needed, the requirements are similar to steel tanks using polystyrene boards except for the method of attachment. If the thermal storage tank is partially buried, the tank wall should be insulated a minimum of 2 ft below the tank backfill. Of course, insulation should not be applied, if the concrete surface is wet, or if the ambient temperature is 50°F and falling.

Hydrostatic and Thermal Performance Testing

Hydrostatic and thermal performance testing for concrete tanks is identical to steel thermal storage tanks.

CHAPTER 11

Thermal Storage Operating and Control Strategies

Introduction

Typically, for an installed thermal storage system, the main goal of the operating and control strategies is to minimize energy costs, to the extent possible (while also maintaining capability to meet peak cooling demand), through reduced peak demand charges (cooling systems); through overall reduced energy use; and through reduced energy use costs compared to the conventional nonthermal storage system. The thermal storage operating and control strategies are, therefore, typically schedule dependent, often based on the local utility rate schedule. As discussed in previous chapters, utility rates sometimes vary depending on the time of day and the season of the year; and the appropriate thermal storage operating strategy must also vary with changing facility needs including varying utility rates. Utility rates may alternatively be based on real-time pricing and a function of the utility system load demand level, and, in those cases, the thermal storage system must be able to respond within an hour, or even 15 minutes, to reduce facility peak demand.

The thermal storage operating and control strategy primarily involves operating and controlling thermal generation or production equipment to achieve the lowest production cost for the capacity provided. With full storage cooling systems, for example, all cooling production equipment must be turned off prior to the start of the on-peak period and not restarted until after the end of the on-peak period. And, on a peak cooling day, the thermal storage charge is depleted shortly after the end of the on-peak period. In addition, for a daily design cycle thermal storage system, the total daily thermal production equipment output must equal total daily facility load, as shown and discussed in Chap. 7. Finally, the thermal storage system operating strategy is often weather and occupancy dependent, and the optimum operation requires some load forecasting abilities.

Per standard practices, as outlined in the ASHRAE handbook, thermal storage operating control strategies can be divided into three key categories:

1. Operating strategies
2. Control strategies
3. Operating modes

The operating strategies provide the overall control methods used to meet the project design intent and provide the "big picture" of how the thermal storage system is designed to operate (e.g., as a full storage system, or as a partial storage system on the peak load day). Control strategies implement operating modes and sequence of operating modes, whereas operating modes describe system means and methods to operate the thermal storage system.

Operating Strategies

Typically, as noted, the main goal of the thermal storage operating strategies is to minimize energy costs, which is achieved through reduced peak demand charges, overall reduced energy use, and through reduced energy use costs compared to the conventional nonthermal storage case. Therefore, the utility rate schedule must be well understood, or, in the case of real-time pricing, continuous pricing feedback must be provided. By shifting when thermal production occurs from when thermal energy is used, as discussed in Chap. 1, energy costs are minimized by the following:

- Minimizing demand charges during the more expensive utility demand rate periods

- Minimizing energy use or consumption during the more expensive utility rate periods

- Operating equipment during more efficient periods (e.g., operating chillers at night with relatively colder condenser water temperatures)

- Operating equipment in a more efficient manner (e.g., avoiding severe low-load operation of chillers and their auxiliary equipment)

A well-designed thermal storage system, therefore, has lower overall energy use compared to a conventional nonthermal storage system, which also help to reduce overall energy costs.

The thermal storage operating strategy provides the overall control strategy used to meet the project design intent. As discussed in previous chapters, the thermal storage system design intent is often to operate either as a partial storage system or as a full storage system, with the partial storage system designed to operate, on the peak thermal load day, with all thermal production equipment (except the backup equipment) operating at full capacity 24 hours per day, whereas the full storage system is designed to operate, on the peak thermal load day, with all thermal production equipment turned off during the utility on-peak period. With cooling systems, the full storage system maximizes the on-peak power demand shift, and, therefore, minimizes the on-peak electric demand and the resultant on-peak electric demand charges.

However, for most facilities, more than 95 percent of the operation will not be on the peak thermal load day, and, therefore, the thermal storage operating strategy, either partial or full storage, for example, may be modified. Regardless of the size of the thermal load, the total daily facility thermal load that must be met by the daily thermal production equipment output. In the case of cooling, as shown in Table 11-1, the partial storage system designed for full-capacity equipment operation around-the-clock on the peak load day can operate as a full storage system on lighter thermal load days. Of course, lighter load days may or may not occur during the season with peak utility rates. Note that on the peak design day, in this example, the facility peak cooling load

Hour ending	Cooling profile (tons)	Partial storage		Cooling profile (tons)	Full storage	
		Chiller output (tons)	TES capacity (ton-hours)		Chiller output (tons)	TES Capacity (ton-hours)
1:00 AM	1,896	4,211	4,541	1,157	3,428	5,916
2:00 AM	1,547	4,211	7,205	944	3,428	8,399
3:00 AM	1,311	4,211	10,106	800	3,428	11,027
4:00 AM	1,202	4,211	13,115	734	3,428	13,720
5:00 AM	1,150	4,211	16,176	702	3,428	16,446
6:00 AM	1,128	4,211	19,260	689	3,428	19,185
7:00 AM	1,266	4,211	22,205	773	3,428	21,840
8:00 AM	1,382	4,211	25,034	844	3,428	24,424
9:00 AM	1,516	4,211	27,730	925	3,428	26,926
10:00 AM	3,279	4,211	28,662	2,002	3,428	28,352
11:00 AM	5,108	4,211	27,765	3,118	3,428	28,662
12:00 PM	5,644	4,211	26,333	3,445	3,428	28,644
1:00 PM	6,383	4,211	24,161	3,896	0	24,748
2:00 PM	6,871	4,211	21,501	4,194	0	20,554
3:00 PM	7,214	4,211	18,499	4,404	0	16,150
4:00 PM	7,521	4,211	15,189	4,591	0	11,559
5:00 PM	7,546	4,211	11,854	4,606	0	6,953
6:00 PM	7,485	4,211	8,581	4,569	0	2,384
7:00 PM	7,213	4,211	5,579	4,403	3,428	1,408
8:00 PM	7,105	4,211	2,685	4,337	3,428	499
9:00 PM	6,432	4,211	465	3,926	3,428	0
10:00 PM	4,676	4,211	0	2,854	3,428	573
11:00 PM	3,927	4,211	284	2,397	3,428	1,604
12:00 AM	2,270	4,211	2,226	1,386	3,428	3,646
Total (ton-hours)	101,072	101,072		61,697	61,697	
Required TES Capacity (ton-hours)			28,662			28,662
Required chiller capacity (tons)		4,211			3,428	

TABLE 11-1   Modified Partial Thermal Storage System Operations

of 7546 tons can be served with a chiller capacity of 4211 tons using a thermal storage system with a minimum capacity of 28,662 ton-hours, and that the total facility cooling load for the day of 101,072 ton-hours equals the total chiller production ton-hours.

However, on a lighter load day with a peak cooling load of 4606 tons, the facility peak cooling load can be served in a full storage mode with a chiller capacity of only

3428 tons using the full thermal storage system capacity of 28,662 ton-hours. And, note also that the total facility cooling load for the day of 61,697 ton-hours equals the total chiller production ton-hours.

In another cooling example, Table 11-2 shows that a full storage system, designed for all thermal production equipment to be off during the utility company's on-peak

Hour ending	Cooling profile (tons)	Full storage		Cooling profile (tons)	Full storage	
		Chiller output (tons)	TES capacity (ton-hours)		Chiller output (tons)	TES capacity (ton-hours)
1:00 AM	1,896	5,615	9,691	948	5,615	9,147
2:00 AM	1,547	5,615	13,760	774	5,615	13,989
3:00 AM	1,311	5,615	18,064	656	5,615	18,948
4:00 AM	1,202	5,615	22,477	601	5,615	23,963
5:00 AM	1,150	5,615	26,942	575	5,615	29,003
6:00 AM	1,128	5,615	31,429	564	5,615	34,054
7:00 AM	1,266	5,615	35,778	633	5,615	39,036
8:00 AM	1,382	5,615	40,011	691	5,615	43,960
9:00 AM	1,516	5,615	44,110	758	0	43,202
10:00 AM	3,279	5,615	46,446	1,640	0	41,563
11:00 AM	5,108	5,615	46,954	2,554	0	39,009
12:00 PM	5,644	5,615	46,925	2,822	0	36,187
1:00 PM	6,383	0	40,542	3,192	0	32,995
2:00 PM	6,871	0	33,671	3,436	0	29,560
3:00 PM	7,214	0	26,457	3,607	0	25,953
4:00 PM	7,521	0	18,936	3,761	0	22,192
5:00 PM	7,546	0	11,390	3,773	0	18,419
6:00 PM	7,485	0	3,905	3,743	0	14,677
7:00 PM	7,213	5,615	2,307	3,607	0	11,070
8:00 PM	7,105	5,615	817	3,553	0	7,517
9:00 PM	6,432	5,615	0	3,216	0	4,301
10:00 PM	4,676	5,615	939	2,338	0	1,963
11:00 PM	3,927	5,615	2,627	1,964	0	0
12:00 AM	2,270	5,615	5,972	1,135	5,615	4,480
Total (ton-hours)	101,072	101,072		50,536	50,536	
Required TES capacity (ton-hours)			46,954			43,960
Required chiller capacity (tons)		5,615			5,615	

TABLE 11-2   Modified Full Thermal Storage System Operations

period, which is a given number of hours (6 hours per day, in this case), at lighter facility loads, can operate with the chillers off for even more hours of the day than is possible on the peak load design day. Note that on the peak design day, in the full storage mode, the facility peak cooling load of 7546 tons can be served with a chiller capacity of 5615 tons using a thermal storage system with a minimum capacity of 46,954 ton-hours. As in the partial storage case, the total facility cooling load for the day of 101,072 ton-hours equals the total chiller production ton-hours. However, on a lighter load day with a peak cooling load of 3773 tons, the facility peak cooling load can be served by operating equipment entirely off peak using the same chiller capacity of 5615 tons, and using the thermal storage system charged to 43,960 ton-hours. Again, note that the total facility cooling load for the day of 50,536 ton-hours equals the total chiller production ton-hours.

Therefore, to minimize energy costs, for both the partial storage case and in the full storage case, the operating strategy should typically be to maximize cooling production during nighttime off-peak hours. If off-peak chiller operation is not enough to meet the day's facility 24-hour cooling load, then next operate chillers at a constant output during mid-peak, as required to meet the remaining facility cooling load. If the added mid-peak operation is still not sufficient, operate chillers at the level (average) load required during the remaining on-peak hours to meet the remaining daily cooling load ton-hours.

Heating thermal storage systems, on the other hand, typically operate in conjunction with when the thermal heat is available. For example, with a solar heating system, the operating strategy is to capture as much heat as possible during daylight hours until the thermal storage system is fully charged, if possible. Another example is cogeneration or combined heat and power (CHP) systems. Cogeneration waste heat should be fully recovered, to the extent possible, whenever the cogeneration system is operating, with the recovered heat either used directly, at that time, or stored in a thermal storage system for later use. Larger cogeneration plants typically operate continuously (except for periodic maintenance) and a heating thermal storage system can be designed as a partial storage system such that the continuous CHP heat output equals the average of the 24-hour heating load. Operating with a properly sized heating thermal storage system, this partial storage strategy maximizes the peak heating load that can be served with a given cogeneration system. Similarly, a base-loaded heat pump system may be designed in conjunction with thermal storage system to increase or maximize the heating loads served.

Example Operating Strategy

The following paragraphs provide an example operating strategy for a cooling thermal storage system:

"The facility's thermal storage system is designed to operate as a 'full storage' system on the peak cooling day. The full storage system operating strategy is to be able to completely turn off the existing central plant chillers during summer utility on-peak periods, up to a facility maximum peak cooling load of 3000 tons and maximum daily cooling load profile of 50,000 ton-hours. With actual lower peak cooling loads and lower daily cooling load profiles, the facility central chiller plant can be in the off-mode for more hours during the day. The summer utility on-peak period is presently from 1:00 to 5:00 PM, Monday through Friday, except holidays. The mid-peak period is from 10:00 AM to 1:00 PM and from 5:00 to 8:00 PM on summer weekdays, and 10:00 AM to 8:00 PM on winter weekdays, except for holidays. Off-peak is all other hours. The summer season commences at midnight on June 1st and continues until midnight on September 30th. The winter season commences at 12:01 AM on October 1st and continues until 12:00 PM on May 31st. Operating control schedule shall match ABC Utility Company rate tariff for the facility.

The overall thermal storage charging or chiller operating strategy is (1) to maximize chiller operation during the off-peak periods and (2) to utilize the mid-peak period for the remainder of the chiller operation, if required, to charge the thermal storage tank fully and to meet the facility cooling load. The table and graphs on the following pages (which would be provided in a real case, but note are not provided in this example operating strategy) depict the peak-cooling-day thermal storage system operating strategy, assuming that facility's central plant chillers are producing 2500 tons each operating in conjunction with a 10,000 ton-hour thermal storage tank. Note that, for each day, the daily chiller production ton-hours must equal the facility's ton-hours."

Control Strategies

Control strategies implement operating modes and sequence of operating modes to fulfill the operating strategy. For example, a typical thermal storage control strategy would be to operate in the "discharge mode" during the utility on-peak period and to switch to the "charging mode" at the end of the utility on-peak period, or when the thermal storage tank is fully discharged. In order to implement the correct control strategy, again, one must know the utility rate schedule (or the current electricity price with real-time pricing systems), the day of the year, the time of day, the expected day's thermal load profile, the status of the thermal storage system charge (e.g., fully charged, 50 percent charged, completely discharged), and know the various available operating modes (described in the following section).

As previously described under operating strategies, control strategies should be modified based on actual facility thermal loads. So, for example, the partial storage system design strategy to have chillers operational at full capacity during the on-peak period to help serve the facility cooling load should be modified, during a lower facility load day, to minimize or eliminate on-peak chiller operation. Therefore, some load forecasting can be helpful and may even be required. One method of load forecasting requires developing a database of facility daily thermal load profiles including the average thermal load for each hour of the day; the daily total ton-hours (in the cooling case) or daily total heat; the maximum daily outside dry-bulb and wet-bulb air temperatures (or minimum temperatures for heating thermal storage systems); and the date. Over the seasons of data collection, the database will be able to provide the minimum, average, and maximum thermal load profiles for different operating conditions such as season, day-of-the-week, and outside-air dry-bulb and wet-bulb temperatures. Given an adequate database, the predicted next day's weather is available from a number of online sources, so a reasonable estimate of the following days thermal load profile can be made, and a thermal storage control strategy plan can be developed.

Control strategies may either be hands-on or hands-off (fully automatic). A hands-off strategy is designed for systems to operate automatically within the general operating conditions. As an example, chillers may operate from 8:00 PM until the thermal storage tank is fully charged. The facility-use-systems operate from 8:00 AM until 10:00 PM at a rate required by load. Such a system can operate automatically without operator intervention, but may not achieve all the savings possible

A hands-on operating strategy may be carefully controlled to suit changing requirements and may start with estimating the next day's cooling load profile. Given the installed cooling production equipment capacity and the installed thermal storage capacity, one can develop a cooling equipment operating plan that maximizes off-peak operation, and minimizes on-peak operation. As previously described under operating strategies, if off-peak chiller operation is not enough to meet the day's facility 24-hour cooling load, then next use mid-peak operation, as required to meet the remaining facility cooling load. If the added mid-peak operation is still not sufficient, operate chillers at the level (average) load required during the remaining on-peak hours to meet the remaining daily cooling load ton-hours.

Using Table 11-1 light load condition as an example control strategy, operate in the "discharge from thermal storage only mode from 12 PM to 6 PM, the charge the thermal storage system and meet the facility thermal load mode from 9 PM to 11 AM, and the discharge from thermal storage and operate chiller equipment to meet the facility thermal load mode from 11 AM to 12 PM and from 6 PM to 9 PM."

Operating Modes

Operating modes describe the thermal storage system operating means and methods and involve charging and discharging the thermal storage system and operating thermal production equipment to charge the thermal storage system and to meet the facility thermal load. Typical thermal storage operating modes are as follows:

- Charge thermal storage only
- Charge the thermal storage system and meet the facility thermal load
- Discharge from thermal storage only
- Discharge from thermal storage and operate thermal production equipment to meet the facility thermal load
- Operate thermal production equipment only to meet the facility thermal load

Other operating mode variations include the following:

- Multiple thermal production equipment choice options (e.g., absorption chiller, steam-turbine-driven chiller, electric-drive chiller)
- Other modes of plant operation (e.g., free cooling, heat recovery)

The following paragraphs provide sample operating mode descriptions for a full storage cooling system example.

Charging Thermal Storage Tank Only

The "Charging Thermal Storage Tank Only" mode typically only occurs if there is no facility thermal load when the thermal storage tank is being charged. In that case, all thermal production goes into thermal storage. However, some thermal storage systems are designed to be isolated from the facility load such that the thermal storage system is charged with separate dedicated thermal production equipment, and any facility load is likewise served with separate thermal production equipment. Also, in accordance with the overall operating strategy, in the "Charging Thermal Storage Tank Only" mode, the thermal production equipment are operated as close to full load as possible until the thermal storage system is fully charged.

Charging Thermal Storage Tank and Serving Facility Cooling Loads

The "Charging Thermal Storage Tank and Serving Facility Cooling Loads" mode is a common operating mode, occurring during lighter load periods of the day, and may occur daily during the mid-peak and off-peak periods. In the "Charging Thermal Storage Tank and Serving Facility Cooling Loads" mode, the thermal production equipment is operated at as close to full load as possible, per the operating and control strategies. Each hour, thermal production equipment output goes both to meet the facility thermal load first, with the remainder of the thermal production equipment output going to charge the thermal storage tank.

Discharging from Thermal Storage Tank Only

In "Discharged from Thermal Storage Tank Only" mode, the entire facility thermal load is met from the thermal storage tank, and the central plant thermal production equipment and associated appurtenant equipment are off (except for distribution pumps). With full storage cooling systems, the "Discharging from Thermal Storage Tank Only" mode occurs daily during the summer on-peak period (and during the winter). Variable-frequency–driven hydronic system pumps draw cooling from the thermal storage system and typically modulate their speed to maintain hydraulically remote building or coil system differential pressures or to maintain coil control valve position slightly less than 100 percent open.

Discharge from Thermal Storage and Operate Thermal Production Equipment to Meet the Facility Thermal Load

In the "Discharge from Thermal Storage and Operate Thermal Production Equipment to Meet the Facility Thermal Load" mode, the facility thermal load is met from both the thermal storage tank and the central plant thermal production equipment. The thermal production equipment serves the facility load at a set, predetermined output and the thermal storage system supplements the thermal demand, with relative demand a function of total thermal production equipment output versus facility thermal load.

Operate Thermal Production Equipment Only to Meet the Facility Thermal Load

The "Operate Thermal Production Equipment Only to Meet the Facility Thermal Load" mode is the conventional nonthermal storage standard operating procedure and occurs when the thermal storage tank is fully charged or fully discharged, or when the thermal storage system is not operational. In the "Operate Thermal Production Equipment Only to Meet the Facility Thermal Load" mode, the central plant thermal production equipment is operated in the conventional fashion with equipment staged or operated to meet the instantaneous facility thermal load.

CHAPTER 12

Commissioning of Thermal Storage Systems

Introduction

The best concepts, design, and construction will not be successful unless systems are operating properly in a manner consistent with the project requirements and design intentions. The main goals of the commissioning process, for new systems, therefore, are to deliver a project that, at the end of construction, meets the owner's project requirements (OPRs), is fully operational and functional, and whose facility staff is fully trained and equipped with the resources they need to properly manage and maintain the facilities, systems, and equipment. Specifically, the commissioning agent seeks to verify that: the systems, equipment, operation, and capacity meet the OPR; the actual operations prove the design sequence of operations; and the operation staff receives proper training. The commissioning agent also helps compile written documentation of the systems, operation and maintenance (O & M) manuals, and system and equipment operation instructions.

Some of the typical objectives of the commissioning process are to:

- Document thermal storage performance objectives in the OPR.
- Verify that the basis-of-design is aligned with thermal storage performance objectives.
- Develop commissioning deliverables (commissioning plan, prefunctional, and functional performance test plans).
- Coordinate commissioning communication between parties (owner, engineers, contractors).
- Accomplish on-going verification that the design intent and construction achieve the OPR.
- Verify and approve that prefunctional equipment checklists have been completed before performing the functional performance tests.
- Witness functional performance tests that document achievement of thermal storage performance objectives before owner acceptance.
- Verify that complete O&M manuals are provided to the owner, which should include complete vendor contact information, installation instructions,

preventive and routine maintenance requirements, troubleshooting guides, cut-away views, wiring diagrams, and part lists, for example.

- Prepare a systems manual assembled from project information provided by engineers, contractors, and equipment or product vendors including consolidated contact information, systems diagrams, sequences of operation, and O&M manuals.
- Verify that maintenance personnel are properly trained.
- Identify operational issues before expiration of warranty and provide the owner with corrective action plans.
- Conduct postoccupancy satisfaction or thermal comfort survey.

To successfully accomplish the above commissioning objectives or scope, the commissioning agent must clearly understand the owner's needs, design objectives, the equipment options, and the factors that are necessary for a successful project. It is recommended that the design engineer be a part of the commissioning team at the onset of the programming phase; a qualified commissioning agent should be involved in the project from development of the design concepts through construction.

This chapter highlights the following:

- Key OPRs for thermal storage systems
- Guidelines of ASHRAE Standard 150, "Method of Testing the Performance of Cool Storage Systems" requirements

Thermal Storage Owner Project Requirements

Different projects and owners, of course, have different project requirements; there are, however, some general key ideas, which apply to almost all projects: a facility that is easy to operate, flexible, expandable, and, also, one that represents a reasonable capital investment. Typical project differences include that some facilities or projects will have an operating staff and others will not. Some facilities will be much more highly automated than other facilities. The role of the commissioning agent may be somewhat different depending on the needs of the owner or facility.

Thermal storage systems are often installed because they represent the lowest life-cycle-cost option. The facility owners (and often the local electric utilities, where they are providing a cash incentive for peak load management) want to know through the commissioning process that they are obtaining the expected thermal storage capacity, demand reduction, and energy cost savings. The owner or facility also wants to be assured that the facility loads are being properly met. Because the facility operation affects the thermal storage operation (just as the thermal storage operation affects the facility's heating or cooling system operations) the thermal storage and the building or facility heating and cooling systems must be tested and their performance documented together. The first step in the commissioning process is to develop or document the OPR, with most of the needed information supplied from the results of the engineering thermal storage feasibility study described in Chap. 8. Ideally, the OPR should be performed before the start of design so that key performance expectations are available to the design engineers at the start of the project engineering phase.

The OPR document should reflect all of the goals or agreements of the various facility stakeholders and can be very extensive in the case of a new building (e.g., documenting civil, architectural, structural, mechanical, power, lighting, building controls, and project, system, or equipment requirements), or can be simpler in the case of a thermal storage system (but still important to document). One key benefit from having the commissioning agent (CxA) facilitate identification of relevant OPR is that often the goals or wishes of some stakeholders are, at least partly, in conflict with the goals and wishes of other stakeholders, and, through the process, a consensus acceptable to (or at least agreed to by) everyone is developed.

The OPR of a thermal storage system typically includes the following:

- Type of thermal storage system to be designed and constructed
- Required thermal storage capacity
- Design chilled water (CHW) or hot water (HW) delta-T for sensible water thermal storage systems
- Expected nominal size and dimensions of the thermal storage tank
- Plan view of the proposed location of the thermal storage tank
- Maximum thermal storage charge and discharge rate
- Thermal storage strategy to be employed (partial storage or full storage)
- Peak day profile showing thermal storage operational strategy
- Geotechnical report documenting maximum allowed soil-bearing pressures
- Instrumentation requirements
- Performance tracking and reporting capability requirements
- Central plant plan, if new generation equipment is being added or modifications to the central plant are anticipated
- Building information and hydronic interface and requirements, as applicable

OPR Design Considerations

For sensible heat water thermal storage systems, the design CHW or HW delta-T (temperature difference between the supply and return temperatures) is, of course, as discussed in previous chapters, a critical piece of information to be provided in the OPR since with a sensible thermal system, the thermal storage capacity is directly proportional to the delta-T ($Q = m*c*$delta-T). Also, as CHW systems operate with relatively smaller delta-Ts, compared to HW systems, as discussed in this book, even a few degrees less than the design delta-T can have a dramatic impact on thermal storage and hydronic system capacity. For example, on a 16° delta-T system, a 1° drop in delta-T to 15° represents a 7 percent loss of thermal storage capacity.

The delta-T of the distribution system and the production system are not necessarily the same as that in the thermal storage tank. In the case of cool storage, the chillers must, however, produce water at least as cold as that stored in the thermal storage tank. The CHW return temperature from the building(s) coils determines how high the CHW return temperature will be in the thermal storage system.

As shown in Chap. 7, the peak day thermal load profile not only provides the facility's peak thermal load, the shape and nature of the load, but also provides the day's required thermal energy production and use requirements as represented by the area under the thermal load profile curve.

For large aboveground and belowground water tanks, a geotechnical report is required and should be performed before the feasibility study, if possible, so that the maximum allowable water height will be known (approximately 62.4 lb/ft^3 multiplied by the height of the water in feet equals the nominal bearing pressure in pounds per square-foot). Otherwise, the feasibility study may have recommended a thermal storage tank size, configuration, or layout that the soil will not support without extensive remediation or requiring an expensive pile or caisson foundation, the costs of which must be included in the feasibility study.

If new generation equipment is planned, the OPR should communicate as much information as possible including equipment catalog cut sheets, equipment selections, equipment views (plan, section, etc.), utility requirements, and equipment schedules. The OPR should also provide a basic plan view of the envisioned central plant equipment layout.

Sometimes, the addition of thermal storage requires modifications to existing buildings connected to the hydronic system; and sometimes existing buildings are being connected to a new hydronic system; or sometimes a whole new district energy system is planned. In those cases, information about existing building systems and existing building loads (or future building loads) used should be included in the OPR.

Basis of Design

The basis of design document is usually prepared by the design engineers and should support and expand upon the OPR. The basis of design allows engineers to document the plans for the upcoming engineering design effort and to communicate those plans to the facility or owner for review and concurrence. In addition to those items outlined in the OPR, the basis of design document also includes code requirements, permit requirements, key points of connection (POC), any construction issues, and any construction phasing requirements.

ASHRAE Standard 150—Method of Testing the Performance of Cool Storage Systems

The purpose of ASHRAE Standard 150 is to "prescribe a uniform set of testing procedures for determining the cooling capacities and efficiencies of cool storage systems." (Note that ASHRAE 94 series standards address the performance testing of individual thermal storage equipment, while Standard 150 addresses the entire storage system.) While this ASHRAE standard is for cooling storage systems, there is also some applicability to a heating storage system. The Standard 150 test serves as an excellent basis for the functional performance test that evaluates whether the project has achieved key OPR performance objectives at the acceptance phase of the project. Standard 150 in brief covers the requirements for the following:

- Testing
- Instrumentation
- Test methods and procedures

- Data and calculations
- Test report

Important aspects of ASHRAE Standard 150 are that it provides details of initialization requirements (e.g., the thermal storage system shall be operated through at least five cycles before testing), the testing apparatus required (e.g., flow and temperature elements), and monitoring points to be measured. Standard 150 also provides instrumentation calibration procedures, and accuracy, precision, and resolution requirements to minimize test uncertainty. For example, temperature difference sensors must have accuracy of at least plus or minus 0.2°F and flow meters must be installed with 20 pipe diameters upstream and 10 diameters downstream (which is always a difficult challenge, if not impossible) in order to achieve an uncertainty of plus or minus 10 percent.

The following procedures are executed to meet ASHRAE Standard 150 requirements:

- Discharge test
- Charge test
- Cool storage capacity test
- Cool storage system efficiency test

One issue that often arises is that rarely will the facility be experiencing the peak cooling load at the time the functional performance tests are conducted. For example, the thermal storage system may be constructed during winter so as to be ready for the summer cooling season. ASHRAE Standard 150 provides some methods of accounting for this common occurrence including operating existing or temporary heat in conditioned spaces to provide a "false" cooling load.

Some of the key CHW thermal storage data to be documented through these tests are:

- CHW thermal storage tank storage capacity
- Thermocline thickness (for stratified thermal storage)
- Diffuser pressure drop
- Tank heat gain
- Load profile achievement (successfully service the facility cooling load)

As discussed in previous chapters, the thermocline is the region where the CHW (or HW) changes temperature between the CHW supply and the CHW return and represents lost, ineffective, or less than fully effective capacity in a stratified thermal storage tank. Typically, temperature sensors located vertically approximately every 2 feet or so within the tank measure and provide the thermal storage tank's temperature profile. Further as discussed in Chap. 9, the diffusers are a critical component of a CHW thermal storage tank, and, if properly designed, allow the CHW to be supplied or withdrawn in a laminar (i.e., nonturbulent) manner to prevent mixing in the thermal storage tank between the CHW supply and CHW return. With respect to CHW pump sizing, the design engineer planned for a maximum pressure drop through the tank and its diffusers, which should be verified during the startup and testing process.

Figure 12-1 shows typical instrumentation requirements and locations. Temperature sensors are required at the inlet and outlet of the thermal storage tank, as well as at the inlet and outlet of the chiller plant. As noted, chiller plant and thermal storage tank inlet and outlet temperatures may be different due to chiller operation and blending or bypassing. Likewise, temperature measurements of CHW supply and return temperatures may also be desirable at the load itself. In addition, as noted, for a stratified CHW thermal storage system, temperature elements will also be needed vertically from the bottom to the top of the tank to determine the percentage of tank charge and the thickness of the thermocline. Flow rate into or from the thermal storage tank can either be measured upstream or downstream of the tank. Note that with a CHW thermal storage system, CHW flows reverse depending on whether the tank is being charged or discharged and a bidirectional flow meter is required. During the charging mode, cold CHW supply is pumped into the bottom of the thermal storage tank and warm CHW

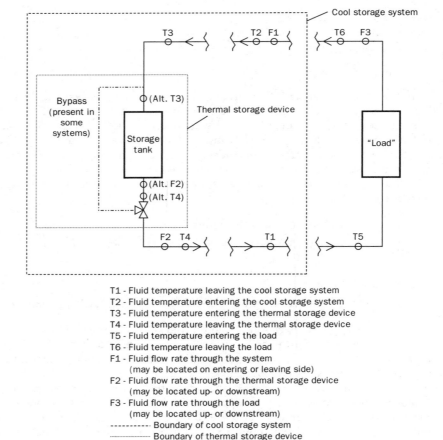

T1 - Fluid temperature leaving the cool storage system
T2 - Fluid temperature entering the cool storage system
T3 - Fluid temperature entering the thermal storage device
T4 - Fluid temperature leaving the thermal storage device
T5 - Fluid temperature entering the load
T6 - Fluid temperature leaving the load
F1 - Fluid flow rate through the system
      (may be located on entering or leaving side)
F2 - Fluid flow rate through the thermal storage device
      (may be located up- or downstream)
F3 - Fluid flow rate through the load
      (may be located up- or downstream)
---------- Boundary of cool storage system
················· Boundary of thermal storage device

FIGURE 12-1 A general cool storage system test schematic showing typical instrumentation requirements and locations. (Reprinted with permission from ASHRAE).

return is drawn from the top of the tank. Conversely, during the discharge mode, cold CHW supply is drawn from the bottom of the thermal storage tank and warm CHW return is pumped to the top of the tank.

Summary

As discussed earlier, the commissioning process should ideally start during the programming phase of a project, with the commissioning authority involved throughout planning, design, construction, and operation (first year), or alternatively, with the design engineer coordinating with the commissioning authority, if that authority begins work during a post-design phase of the project. A key first step is documenting the OPRs. The commissioning authority also prepares the prefunctional and functional performance test plans. ASHRAE Standard 150 provides a uniform testing procedure to evaluate the performance of an installed thermal storage system. Success of the project is proven when the OPRs are being met. Ideally, the commissioning agent verifies annually or per an ongoing schedule that the system continues to be operated and functions in accordance with the OPR.

CHAPTER 13

Sustainable Thermal Storage Operations

Introduction

As described in the preceding chapters, various factors and requirements work together to help achieve sustainable thermal storage operations. Facility management and facility plant operators or maintenance personnel with thermal storage or those planning to add thermal storage, as well as thermal storage engineers, thermal storage vendors, and other thermal storage stakeholders should understand the following: thermal storage basics; basic thermal storage economics; thermal storage plants; thermal storage plant systems; thermal storage plant equipment; thermal storage operating strategies; utility rate structures, in general, and local utility rate structures applicable to a specific location or project; and general, proper plant operations and maintenance requirements. A basic knowledge (at least) of plant engineering and operations as well as district energy systems is helpful (e.g., the purpose, basic configuration, and operation of central chiller, boiler, or CHP plants, thermal production equipment, pumping systems, heat transfer systems, thermal distribution systems, and thermal use systems, such as air-handling coils or turbine inlet cooling coils).

As the old adage states, "one cannot manage what one does not measure," sustainable thermal storage operations should have comprehensive metering and monitoring system as part of the plant or facility control system. With good plant metering and monitoring, production quantities and auxiliary or parasitic loads can be measured, and various metrics can be calculated and trended and benchmarked, thereby, providing key tools for performance diagnostics; optimizing operations; and maintaining optimized operations.

Managers of sustainable thermal storage systems value their plant operators, invest in their training, and have mechanisms in place to facilitate feedback and good, open communication to improve plant operations and performance. Further, thermal storage plant–operating strategies and their requirements, consequences, or costs should be fully analyzed and the results understood by all concerned stakeholders, especially plant operators. Given the net production quantities and the total auxiliary and parasitic loads determined from plant metering, and given (or calculating) the total cost of production for each plant produced utility, the unit production costs can also be determined and benchmarked. Finally, similar to any plant facility, financial resources need to be obtained and expended to maintain the long-term sustainability of the thermal storage plant.

Understanding the Thermal Storage Plant

Understanding thermal storage plant operations begins, of course, by understanding basic thermal storage concepts and basic thermal storage economics, as previously discussed in this book. Taking a bottom-to-top approach, understanding a particular thermal storage plant begins by understanding the function and construction of each individual piece of major plant equipment. Thermal storage plant personnel (as well as thermal storage engineers) should be able to discuss how individual pieces of equipment, and the main components that make up that equipment, function. Various pieces of plant equipment are piped or connected together to make plant systems [e.g., chilled water (CHW) system, condenser water system, hot water (HW) system, steam or condensate systems], and a single piece of equipment can be part of multiple systems (e.g., a steam-driven centrifugal chiller). Thermal storage plant personnel should do the following:

- Trace out all plant systems (i.e., follow the piping through the plant from start to finish).

- Be able to draw, from memory, a schematic diagram of each plant system showing all major equipment and key system components including major isolation, safety, and control valves.

- For each plant system, be able to discuss each piece of equipment in the system, and how the equipment functions as part of the system.

- Be able to discuss how plant systems function efficiently as part of the overall thermal storage plant.

The various thermal storage plant systems combine to form a working thermal storage plant that typically has various equipment options to choose from and various plant-control strategies. The cost of equipment or system operating choices can be analyzed and understood and thermal storage plant personnel should have this information to make good economical decisions (as a hypothetical example, only operate the absorption chiller during the summer on-peak and mid-peak utility rate periods, and maximize use of the electric-drive centrifugal chillers during the off-peak hours).

Sustainable on-site thermal storage operations require a team effort by facility or plant management, plant operations, and often by outside consultants (engineering, energy-purchase, financial) with each team member playing an important role. While each team member has a role to play and a certain level of expertise, all thermal storage plant team members should be familiar with all the basic goals of the thermal storage plant.

Typical goals of the sustainable thermal storage plant include the following: to provide the thermal capacity to meet peak facility thermal loads with much lower installed equipment capacity; to minimize energy use and demand costs; and to maximize the return-on-investment (ROI) in the thermal storage plant itself to pay back investments, to fund plant operations and maintenance, and, hopefully, to provide for a reserve fund for future equipment replacement. The ROI is maximized, typically, when total energy costs are minimized, and by maintaining a high plant availability (i.e., keeping the plant up and running). A high plant availability results from good operation and maintenance procedures or practices but is also very much a function of equipment quality and plant design.

Specifically, the ROI is maximized by minimizing plant purchased utility costs, with electricity costs often being a major cost driver. Depending upon the location of the thermal storage plant, electricity purchase options (and associated cost ramifications) vary from the only option of buying electricity from the local utility to buying electricity via the real-time spot market and to generating some or all of a facility's power requirements. Some facilities have a full-time energy manager to work on the above issues, while other facilities hire outside consultants for professional recommendations. Note that electricity costs are also minimized by maximizing plant-operating efficiencies, which minimize energy use per unit of production (e.g., kW/ton), and is affected by various factors as previously discussed in this book.

Maximizing the ROI involves properly metering and sometimes billing for thermal storage plant–generated services. Therefore, thermal storage plant management and staff and consultants need to understand the basics of the different types and functions of meters and how the information on those meters is translated onto energy usage reports and customer utility bills (as applicable), which is beyond the scope of this book.

Thermal Storage Data Gathering

As noted, one of the critical requirements for a sustainable thermal storage plant is the ability to gather sufficient information or data in order (1) to meter and bill for thermal storage services properly and (2) to monitor and trend performance to receive feedback and to help maximize performance and minimize costs. The first step begins with data gathering, which can be automated as part of the plant control system.

Metering

Thermal storage systems are sometimes part of a district energy system, and, in many cases, customers are billed for their electricity and thermal usage. Meters are typically used to measure the facility or building usage of electricity, CHW, heating HW, steam or condensate, and domestic HW, for example. Measurement of electricity should include both a measurement for consumption and for demand. Electricity meters are not only required for each individual building, in the case of a district energy system, but are also required to capture major equipment usage, plant auxiliary loads or consumption and parasitic losses for individual systems and equipment (e.g., chiller power to measure kW/ton). Many types of fluid meters need to be located with the proper upstream and downstream straight-run diameters to record flow with reasonable accuracy. Where steam is supplied to a customer, in addition to steam flow meters, condensate meters should also be installed, and customers charged on their net BTU consumption obtained by computing the total energy consumed equal to the energy (enthalpy) supplied in the steam minus the energy returned in the condensate. In this way, customers get a credit for condensate returned, and condensate that is not returned to the plant is accounted for and the customer has an incentive to fix condensate leaks and repair insulation.

Finally, significant time and costs can be saved by employing meters now available in the marketplace, which have significant technology enhancements compared to meters in the past. For example, new smart meters can use wireless technology that allows meters to be read remotely, and the output transferred directly (without separate human data entry) to automated billing systems and energy report generation. Smart meters can notify facilities when measurements fall out of expected ranges and when there is a problem or even impending problem with the meter.

Monitoring

As outlined in the following section, Thermal Storage Data Analysis, in addition to metering the thermal storage plant utility production and the various equipment or system power usages (or fuel usages), the plant control system also needs to monitor and record total thermal usage (e.g., CHW/HW/steam). In addition, the monitoring and control system should also monitor and record all of the key plant flows, temperatures, and pressures. When any parameter is out of the acceptable range, an alarm should be generated by the monitoring and control system. As discussed, the plant monitoring data should be used to compute equipment and system efficiencies and metrics for use by plant operators and management to better track and trend thermal storage operations, to help with plant troubleshooting, and to provide thermal storage plant optimization feedback.

Thermal Storage Data Analysis

Given the plant data, plant personnel or outside consultants can analyze thermal storage plant operations. Important information can be gleaned directly from the raw data (which may include some basic calculations such as thermal load from flow rate and temperature difference), and also a number of metrics that can be calculated to analyze thermal storage plant operations. Results of data analysis can be benchmarked against other facilities and other plant-operating options to draw comparisons, contrasts, and conclusions.

Review of raw data obtained directly provides important information regarding, for example:

- Total plant-generated cooling, over a given period (e.g., day, month, year)
- Total plant-generated heat, over a given period
- Facility thermal load at any point in time including maximums and minimums (i.e., peak facility load and base facility load)
- Actual operating profile of thermal load versus time (e.g., the daily facility 24-hour cooling load profile versus chiller plant operation)
- Total electricity costs for a given period (typically monthly or annually)
- Total electricity consumed by the plant, for each process (e.g., cooling, or heating), over a given period
- Total electricity consumed by each major piece of equipment (note that this information is many times not available; however, if available, it allows for a more detailed analysis: equipment level versus overall thermal production plant level)
- Total fuel consumed for any fuel-burning devices (e.g., direct-fired absorption chillers, or backup HW boilers)

- Key plant-operating parameters including temperatures, flows, and pressures of:
  - CHW supply and return
  - HW supply and return
  - Condenser water supply and return
  - Steam and condensate, if applicable
  - Outside dry-bulb and wet-bulb (temperatures)
- Thermal (heating and cooling) deliveries or sales for a given period (total energy units and accounts receivable)

Metrics

In addition to the raw thermal storage or facility data, important metrics (performance indicators) can be developed or calculated from the raw thermal storage plant data to better understand thermal storage plant operations and can provide guidance toward more efficient sustained thermal storage plant operations. Care should be taken not to draw too broad of a conclusion or to make false assumptions regarding individual metrics. Key thermal storage metrics include the following:

- Thermal storage tank efficiency (thermal storage discharge capacity divided by thermal storage charge capacity)
- The amount of energy consumed per unit of cooling produced (e.g., kW/ton)
- The coefficient of performance (COP), which is equal to the cooling or heating achieved, divided by work or energy required in consistent units, and, in the cooling case, is equal to the inverse of the kW/ton in consistent units
- Heating plant efficiency (heat out divided by energy in)
- Cost per unit of cooling generated for a given period including for different utility tariff rate periods (e.g., dollars per ton-hour)
- Cost per unit of heating generated for a given period (e.g., dollars per million Btu)
- Cost to produce cooling with thermal storage facility versus the business as usual (BAU) case without thermal storage
- Amount of money saved by employing thermal storage versus the conventional, nonthermal storage BAU case
- Return on investment (ROI) or internal rate of return (IRR)

To compare the amount of money saved by employing thermal storage versus the conventional BAU case, the cost to produce individual applicable facility services (e.g., CHW, heating HW, steam, etc.) must be determined on a per unit delivered basis (e.g., ton-hou, million Btu, therm, etc.) versus the BAU case. As discussed, utilities provided by the plant should be fully metered and all production, operating, or maintenance costs accounted for. The unit cost is equal to the total cost to generate that utility (e.g., CHW) divided by the net production amount.

Determining the total cost of individual facility-provided utility services may sometimes be challenging and the results skewed depending upon how costs are allocated. For example, how to account for labor costs between the various plant-supplied utilities is an important question, because not all equipment requires equal supervision. For example, a high-pressure boiler probably has mandated 24-hour-per-day licensed operator requirements, while an electric-drive centrifugal chiller with a unit control panel may only need to be checked periodically. Comparing the unit cost to provide plant-generated utilities versus the BAU case should show that a positive rate of return is being achieved, that is, the cost to generate utilities by the facilities, on a unit cost basis, should be less than the BAU case.

As previously discussed, most purchased utilities have some time-of-use or tiered consumption rate schedule that must be factored into the unit cost calculations and analysis of the thermal storage plant. For example, the cost to generate CHW will likely be less expensive at night versus during the day (unless the facility is on a flat tariff rate schedule), while the value of CHW thermal storage output will likely be more valuable during the day than at night.

The total net amount of money saved, by employing thermal storage versus the conventional BAU case without thermal storage, can offer a realistic estimate of the financial performance of the thermal storage plant and is calculated by estimating the operating cost to generate and meet thermal loads in a conventional, nonthermal storage manner without thermal storage versus with the use of thermal storage.

Benchmarking

Benchmarking can be used to help analyze the results of the above-described metrics, and can allow thermal storage plant personnel to compare their thermal storage plant cost of production with other facilities. An example of a common metric that is often benchmarked is the cost to produce CHW. Another common example of a benchmarked value is facility energy use per unit floor area, and the results can depend on the type of facility, type of facility construction, facility location, season, weather, and occupancy schedule, and how the CHW is produced. Therefore, benchmarking can sometimes be misleading. For example, a thermal storage facility with a favorable low energy use per square foot metric compared with other thermal storage facilities with higher values may be the result of a more benign climate compared with that of others located in more extreme climates and the result of shorter operating hours than other facilities in the thermal storage plant group being compared against.

Billing

For those owner-operators who charge customers for thermal storage plant–generated utilities, accurate and clear billing is important in sustaining efficient thermal storage operations as it may provide the funding (or a portion of the funding) for operations, for maintenance, and for the reserve funds necessary for future equipment replacement. Billing best practices are those practices that are as follows:

- Transparent
- Auditable
- Fair (reasonable)
- Easy to follow
- Follow generally accepted accounting principles (GAAP)
- Account for all costs (including capital or debt costs)

Following these best practices, all costs associated with producing CHW, heating water, power, and steam, for example, are appropriately accounted for and assigned, including, for example:

- Electricity purchases
- Fuel purchases
- Operating staff
- Maintenance
- Administration
- Water
- Chemicals
- Supplies
- Debt service
- Depreciation and reserve fund

Whenever possible, cost should be assigned to the applicable individual utility service. For example, work on a chiller should be assigned to the cost to produce cooling, while work on a boiler or heat recovery unit should be assigned to the cost to produce heating. If a cost cannot be allocated solely to a given thermal production, the cost should be allocated appropriately between utilities. Different methods exist to allocate costs properly among related utilities. One method is simply to divide up the costs equally between the different thermal production outputs. However, other cost splits may have more merit including dividing the costs proportionally to match the actual proportional value of the utilities.

Auxiliary and parasitic loads should also be computed to determine total consumption versus net output. Examples of auxiliary and parasitic loads include the following:

- Plant pumps
- Cooling tower fans
- Forced draft blowers
- Compressors
- Water treatment systems
- Deaerating (DA) tank (steam)

With operating costs and losses known, appropriate billing rates can be developed. The bill itself should be easy for consumers to follow and should include relevant metrics to indicate efficiency such as Btu-per-square foot, and utility usage for the same time period in previous years.

Control Strategies

Thermal storage-operating and control strategies are discussed in Chap. 11, and the number and type of possible control strategies usually depends upon the thermal storage plant size (storage and generation) versus actual facility thermal loads; and the nature or size and type of available equipment options. Equipment-operating strategies can be determined from the unit cost analysis previously described, and a matrix can be developed to show all plant equipment options listing each and every equipment or system and preferred choices. The matrix should show the available thermal production equipment, thermal storage, and the marginal operating costs, and even relative value or cost (e.g., on-peak operational savings or cost) to help plant operators determine good equipment choices or operating strategies.

However, as important, as previously discussed, the thermal storage systems are typically sized for the peak day load. Therefore, for a majority of the year, the facility thermal load is less than peak thermal load, and, hence, the thermal storage-operating strategy can and should be modified to maximize possible energy savings. As shown in Tables 11-1 and 11-2, energy savings are maximized by maximizing charging of the thermal storage system during lower cost off-peak utility rate periods and by minimizing or eliminating thermal production during higher cost on-peak and mid-peak utility rate periods.

Note that a modern, technologically advanced, robust, fast-acting, control system capable of calculations and automated decision making is very helpful, if not essential, in implementing various operating strategies.

To compare control strategies, financial pro forma models can also be created to calculate or estimate total and marginal cost, value, and amount of money saved for each piece of equipment or system operated. Cost and values will, of course, as discussed, be affected by utility time-of-use rates, seasonal differences in utility costs, as well as by the cost of fuel (which can change depending upon season or purchase agreements or terms), all of which need to be factored into account in order to better understand equipment or system-operating choices and their resultant cost or revenue implications or consequences.

Equipment options or system operations are generally selected based on minimizing energy costs. For example, at a given time, a choice might need to be made on whether to operate a steam absorption chiller or an electric-drive chiller, and a matrix, as described above, can provide answers to what is the lowest cost equipment to operate. In this case, for example, it may be cost effective to operate the absorption chiller during the on-peak period but not during the off-peak period. In addition, with a thermal storage plant, there may be times when a use for the recover of waste heat overrides or changes cost calculations.

Another operating strategy is to minimize parasitic and distribution losses, and to maximize thermal storage plant efficiencies. As with any energy project, reducing waste is step 1, minimizing facility loads is step 2 (daylighting, more efficient lighting, building insulation, more efficient windows, etc.), and minimizing thermal storage plant losses, where possible, should be studied, reviewed, and implemented on an on-going basis as step 3. A common loss occurs in the condensate system where condensate is not returned from buildings to the plant and where heat losses occur in condensate piping, losing energy. Another common example is poorly maintained steam traps that often leak by wasting energy (the steam traps let steam pass through to the condensate system). All piping should be well maintained, well insulated, and free from leaks. Plant pumps should be selected and operated to minimize pumping horsepower. Most importantly, pumping horsepower can be minimized by maximizing hydronic system delta-Ts. Of course, many losses are set by the design and construction of the thermal storage plant, for example, pipe sizes, and inherent pressure drops.

Improving thermal storage plant including chiller plant efficiencies is an important and interesting subject worthy of a separate chapter or even a separate book by itself. The challenge with improving thermal storage plant efficiencies is that equipment or systems are interrelated and the operations of one system, for example, affects the operations of another system. For example, higher flow, colder condenser water allows electric motor–drive chillers to operate more efficiently requiring less motor horsepower. But providing higher condenser water flows requires higher condenser water pump power (for a given system), and providing colder condenser water requires higher cooling tower fan horsepower (for a given wet-bulb temperature). The question is whether the chiller horsepower savings are more than the condenser water pump and fan horsepower increases (assumes variable-frequency-drive motors), and the answer will depend on equipment part load performance. As another example, the cost of operating a chiller to generate CHW for turbine inlet cooling may be far outweighed by the value of increased combustion turbine generator output. Algorithms exist to optimize individual plant systems, such as the CHW and condenser water systems, and to operate equipment along its natural curve of best efficiency points, for a given load and operating condition. Using power consumption meters, empirical method can also be used to map power consumption versus various applicable operating parameters to determine operating conditions that minimize power consumption.

Operator Training

Plant operators are very important in the success of any facility operations, as it is the plant operators who are on the "front line" and who can observe and report plant-operating conditions and make key suggestions for improvements. And it is the plant operators, for example, who implement and make work (or not) management or consultant-recommended plant energy conservation measures. Having knowledgeable, trained plant operators is important for thermal storage plants, and, therefore, on-going operator training is important to sustainability. No matter how much an individual knows, much more can be learned. Additionally, with the rapid advance of technology, continuous training or education is required just to maintain proficiency. Also, rules and regulations change, and facility operators need to be familiar with any regulatory items affecting their facility. Changing priorities may also require education, as energy efficiency and pollution reduction becomes more important to a facility's overall mission for example.

Maintenance

Thorough on-going maintenance is essential to sustainable thermal storage plant operations. Every plant must have a preventive maintenance (PM) system, and planned maintenance shutdowns must typically be carefully coordinated. If the thermal storage plant was designed with back-up equipment, and the facility thermal load is low and the thermal storage system is charged, unexpected failures may be successfully handled without disruptions to operations. Unplanned disruptions to thermal storage plant operations can be minimized with good operations and maintenance, by operating the plant in a stable and safe manner, and by maintaining equipment in good working condition. A number of software programs and systems are available to manage maintenance operations, with different degrees of sophistication. However, the basic concept is to track each and every piece of equipment, heat exchanger, valve, control device, and every other device needing maintenance, along with the corresponding maintenance items required for each item (e.g., change fluids, lubricate fittings, change belts, check clearances) as well as the schedule for those maintenance items (e.g., daily, monthly, quarterly, annually) to develop an overall plant maintenance schedule.

Reserve Funds

All equipment, no matter how well maintained, wears out eventually and must be replaced. Facilities can and should plan for equipment replacements by establishing a reserve fund. A thermal storage plant reserve fund is another equipment matrix that lists each and every piece of equipment along with the equipment's respective:

- Date put into service
- Remaining life expectancy in years (or date to be replaced)
- Current cost to replace
- Estimated cost to replace in future
- Current reserve amount
- Calculated current reserve requirement
- Difference between the current reserve amount and the calculated current reserve requirement
- Annual amount needed to be added to the reserve fund

The future cost to replace is estimated by escalating the current cost to replace by the number of years of equipment life remaining. The calculated current reserve requirement is determined from the future cost to replace and the remaining equipment life, taking into consideration the time value of money. As a simple example, neglecting inflation and savings interest, if a piece of equipment costs $100,000 and the life expectancy is 20 years, the facility should add $5000 to the reserve fund every year, and, if 6 years of life remained, then the reserve fund should hold $70,000 for that piece of equipment. Of course, in reality, purchase cost escalation as well as interest on reserve fund investments must be taken into account.

Let People Know the Great Results of Thermal Storage

Finally, let people know the great results of having a sustainable on-site thermal storage system. Let people know that on-site thermal storage is a time-tested, proven technology that offers many important benefits to building and facility owners and operators, to local and regional utility systems, to a country's economic competitiveness and security, and to human society as a whole. Let people know that sustainable on-site thermal storage's important benefits include the following:

- Lower facility energy costs
- Reduced peak demand
- Reduced energy consumption
- Improved facility plant efficiencies
- Smaller thermal production equipment requirements

- More flexible plant operations
- Added backup capacity
- Lower facility plant maintenance costs
- Less strain on infrastructure
- Potential lower capital costs versus a conventional, nonthermal storage system
- Overall economic benefits
- Environmental benefits
- Benefits to the local utility company
- Benefits to the society

CHAPTER 14

Chilled Water Thermal Storage Case Study

University Hospital

A chilled water (CHW) thermal energy storage (TES) study was conducted for a local university. The university includes a growing educational health-science institute, founded in the early 1900s. At the time this case study was conducted, the university had an enrollment of more than 3000 students from 80 countries and nearly all 50 states. The academic programs feature over 100 degree and certificate programs in the areas of Medicine, Dentistry, Nursing, Public Health, Allied Health as well as graduate studies. In addition to the teaching and research functions associated with the general campus, the university also maintains a hospital, children's hospital, and emergency room.

The university is located at an elevation of approximately 1300 feet (ft) above sea level. The university enjoys a desert climate with sunny days virtually year round. During the winter, the nighttime temperature occasionally falls below the freezing point of water. On summer days, the temperature can reach more than 100°F.

Study Purpose

The university commissioned a Chiller Plant Expansion or TES Study to determine the best method of meeting the growing demand for central plant cooling on the university campus.

Similar to many other institutions, the university continues to expand and therefore campus loads (cooling, heating, and electrical) continue to grow. When this study was conducted, peak electricity usage had surpassed the capacity of the existing central cogeneration plant, and the university was forced to purchase power from the utility company during most daytime periods, including the on-peak period (the most expensive electric utility time period). The electric demand resulting from the operation

---

This chapter provides a simple example of a chilled water thermal storage study, not including the requisite drawings and appendices, which include energy calculations and budget cost estimates. Chapter 15 provides a simple example of an ice thermal storage study.

of the existing electric-drive chillers and ancillary equipment within the central plant was a significant portion of campus demand (about 25 percent). The university realized that TES could provide a method of shifting cooling production from the on-peak period to the off- and mid-peak periods, reducing the amount of power that would need to be purchased and instead allowing the cogeneration plant to meet facility electric load more effectively. The university also realized that an additional benefit of TES would be in providing a means for expanded cooling capacity and wanted to determine the most cost-effective method of providing for the existing and future cooling loads.

Campus Overview

The main campus of the university comprises approximately 50 buildings with gross areas ranging from hundreds of square feet to hundreds of thousands of square feet. The buildings consisted of academic classrooms, laboratories, dormitories, administration centers, libraries, and a hospital. When this study was conducted, 24 buildings were served by the CHW system from the central plant.

CHW is pumped from the central plant through an underground tunnel to each of the buildings, where the CHW is routed to the cooling coils of air-handling units or fan coil units. In some buildings, piping systems are directly connected to the CHW distribution system of the campus, while other buildings use secondary pumps to boost the CHW flow through the buildings (Table 14-1). Table 14-1 provides the building list an estimated peak cooling loads.

Existing Cooling and Heating Systems

The central plant consists of three areas: the original central heating and cooling plant, an expanded chiller plant, and a cogeneration facility. The three plants are adjacent to each other.

The original central heating and cooling plant is located on the east side of the Power Plant. In addition to office and other administrative space, the approximately 10,000 square-foot (ft^2) building housed three absorption chillers with a total cooling capacity of approximately 3100 tons, two 2500 gallon per minute (gpm) primary CHW pumps, one 3250 gpm primary CHW pump, and three 5000 gpm condenser water pumps when this study was conducted. A three-cell, 15,000 gpm, ceramic counter flow cooling tower was located outside the plant at ground level (Fig. 14-1). Figure 14-1 shows the old absorption chillers.

The university built the expanded chiller plant, located on the west side of the Power Plant, in the late 1980s as an expansion of the original heating and cooling plant. At the time of this study, the approximately 6000 ft^2 building housed five electric-drive centrifugal chillers with a total cooling capacity of 5750 tons, three 4000 gpm CHW pumps, and three 5000 gpm condenser water pumps. One five-cell cross-flow stainless steel cooling tower was located on the roof and could provide 15,000 gpm at 85°F (14°F delta-T), which was insufficient to provide condenser water to all five centrifugal chillers.

The cogeneration plant consisted of two 5-megawatt (MW) combustion turbine generators (CTGs) with duct-firing and heat recovery steam generators (HRSGs) when

Building name	Area (ft^2)	Cooling load factor (sf/ton)	Cooling load (tons)	Gpm@ 12°F delta-T
Academic building	106,400	240	443	887
Academic building	96,123	240	401	801
Academic building	64,880	240	270	541
Academic building	40,322	240	168	336
Academic building	29,038	240	121	242
Academic building	21,400	240	89	178
Academic building	20,485	240	85	171
Academic building	10,408	240	43	87
Business center	30,242	240	126	252
Cafeteria	9,504	240	40	79
Campus church	36,171	240	151	302
Convenience center	84,958	240	354	708
Library	85,493	240	356	713
Market	31,900	240	133	266
Medical center				
Emergency room	150,700	188	802	1,604
Hospital	555,000	188	2,952	5,907
E wing	53,000	188	282	564
Nursing tower	50,000	188	266	532
Radiological facility	9,800	188	52	104
W wing	422,000	188	2,245	4,491
Medical offices	214,000	240	892	1,784
Outpatient clinic	100,000	200	500	1,000
Power plant	22,945	240	96	191
Recreation center	95,000	240	396	797
Research institute	139,733	200	699	1,393
Student center	22,888	240	95	191
Student housing	63,811	240	266	537
Student housing	51,070	240	213	426
University offices	7,373	240	31	63
Total	**2,624,644**		**12,567***	**25,147***

* Nondiversified.

TABLE 14-1 Building List and Estimated Peak Cooling Loads

FIGURE 14-1 Absorption chillers ABS-1 and ABS-2.

this study was conducted. There was also a steam turbine generator (STG) capable of producing approximately 1.2 MW. Table 14-2 provides a summary of the existing chillers before the central plant retrofit.

The total existing installed cooling capacity was approximately 8850 tons. Due to CHW pumping constraints, the total CHW plant capacity was limited to approximately 7600 tons.

Existing Energy Costs

In the early 2000s, state residents and businesses experienced what newspapers and politicians labeled an "Energy Crisis." Spot electricity prices escalated from around $30 per megawatt-hour (MWH) to more than $300 per MWH, a 10-fold increase. Since then, spot electricity prices returned to levels comparable to those seen before the electricity price escalation, but the volatility in energy prices still made difficult times for those with significant energy budgets. Even though the university had its own cogeneration plant that generated approximately 80 percent of annual campus kilowatt-hours (KWhs), the capacity of the cogeneration facility could not meet the peak electrical demands of the campus most of the year. As a result, the university had to purchase electricity from the local utility company to meet its electrical demands. The escalating energy prices and ever-increasing energy use were important reasons for the university to decide on exploring energy cost-saving options including the installation of a TES system.

Location	Central heating and cooling plant			Chiller plant				
Tag	ABS-1	ABS-2	ABS-3	CH-1	CH-2	CH-3	CH-4	CH-5
Equipment type	Absorption	Absorption	Absorption	Centrifugal	Centrifugal	Centrifugal	Centrifugal	Centrifugal
Manufacturer	Trane	Trane	Trane	Trane	York	York	Trane	Trane
Cooling capacity	719	719	1,709	1,250	1,000	950	1,250	1,250
kW/ton	–	–	–	0.57	0.54	0.56	0.57	0.57
Steam rate (lb/ton-hour)	19	19	19	–	–	–	–	–
Chilled water flow rate (gpm)	1,438	1,438	4,100	3,333	2,400	2,533	3,333	3,333
Chilled water EWT (°F)	54	54	55	51	55	51	51	51
Chilled water LWT (°F)	42	42	45	42	45	42	42	42
Condenser water flow rate (gpm)	2,700	2,700	6,000	3,874	2,765	2,939	3,874	3,874
Condenser water EWT (°F)	85	85	85	84	80	84	84	84
Condenser water LWT (°F)	101	101	102	93	90	93	93	93
Refrigerant type	LiBr	LiBr	LiBr	R-123	R-500	R-134	R-11	R-11
kW	10.4	10.4	29	713	535	535	713	713
Voltage	460	460	460	4,160	4,160	4,160	4,160	4,160
Phase/Hz	3/60	3/60	3/60	3/60	3/60	3/60	3/60	3/60

TABLE **14-2** Summary of Existing Chillers Before Retrofit

Electricity Rate Schedule

The local utility company generally charges large electricity users, including the university, based on a time of use (TOU) rate schedule. The TOU rate schedule consists of two main charges: the demand charge and the energy charge. The demand charge is based on peak power usage in a given time period while the energy charge is based on the actual electrical energy consumed. Both rates vary with different time periods of the day (on-peak, mid-peak, and off-peak) and different seasons of the year (winter and summer). A standby rate is also charged to energy users such as the university, with their own cogeneration facilities. Table 14-3 provides the utility rate schedule at the time of the study.

The winter season is from October to June, and the summer season is from June to October. The on-peak time period is on weekdays from Noon to 6:00 PM during the summer season only. During the summer season, the mid-peak time period is on weekdays from 8:00 AM to Noon and from 6:00 to 11:00 PM, while the off-peak time period is on weekdays from 11:00 PM to 8:00 AM and on weekends. During the winter season, the mid-peak time period is on weekdays from 8:00 AM to 9:00 PM, while the off-peak time period is on weekdays from 9:00 PM to 8:00 AM and on weekends.

Existing Central Plant Cooling Load

Based on information provided by the university, an estimated 24-hour cooling load profile was developed for the peak day. As shown in Fig.14-2, the existing peak cooling load was approximately 7200 tons and occurred around 5:00 PM. The minimum load was approximately 1500 tons, and occurred around 5:00 AM. The total peak day cooling load was approximately 100,000 ton-hours.

	Summer	Winter
Customer charge	$391.46	$391.46
Demand charge (to be added to customer charge):		
Facilities-related component:		
All kW of billing demand, per kW	$1.58	$1.58
Time-related component:		
All kW of on-peak billing demand, per kW	$13.49	N/A
All kW of mid-peak billing demand, per kW	$1.88	$0.00
All kW of off-peak billing demand, per kW	$0.00	$0.00
Standby charge		
All kW of standby demand	$0.69	$0.69
Generation standby charge		
All kW of standby demand	$0.29	$0.29

TABLE 14-3   TOU Rate Schedule at Time of Study

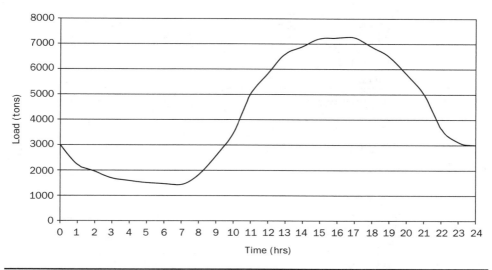

FIGURE 14-2 Existing peak day cooling load profile.

Future Cooling Load

The university estimated the additional future CHW cooling load required to be approximately 920 tons. This future cooling load was estimated to have linear annual growth, distributed across 8 years. Additionally, the university planned to add a 250,000 ft² academic complex, constructed in two phases, with a cooling load of approximately 1200 tons. Table 14-4 provides the estimated future university cooling load.

By incorporating these future cooling loads, an estimated future 24-hour cooling load profile was developed for the peak day, as shown in Fig. 14-3. The estimated future peak-cooling load was determined to be approximately 9300 tons. The minimum cooling load on the future peak day was estimated at approximately 1800 tons. The total cooling required for the day was projected to be approximately 130,000 ton-hours.

Central Chiller Plant Alternatives

Three central chiller plant alternatives were generated to meet the future cooling needs of the university: a conventional chiller plant with no TES system, a full storage TES plant, and a partial storage TES plant. The primary considerations applied in the overall evaluation of proposed alternatives were the ability of the system to reliably meet the cooling loads, initial capital cost of the systems, ease of installation and available space, impact on energy and operations costs, ease of maintenance, and ability of the potential system to work with existing systems. The optimum solution determined in the evaluation was not necessarily the solution with the lowest first cost, the system which provides the highest energy cost savings, or any other single factor but the solution which provided the best combination of factors as evaluated on a life-cycle-cost basis.

Phase	1	2	3	4	Total
Additional cooling load (tons)	97	134	485	206	922
Academic complex					1,200
Total					**2,122**

TABLE 14-4   Estimated Future Additional Cooling Load

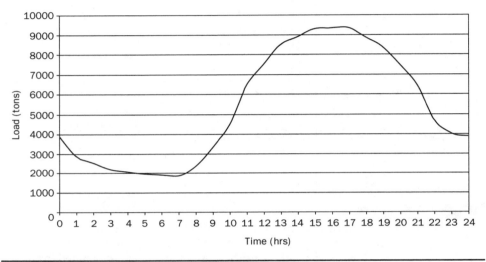

FIGURE 14-3   Future peak day cooling load profile.

Alternative 1—Conventional Chiller Plant (Base Case)

Alternative 1 considered using the existing centrifugal chillers and adding new centrifugal chillers, and operating the plant as a conventional chiller plant (i.e., adjusting the chillers to meet the projected campus cooling load), which was how the plant was already operating. The combined capacity of the existing centrifugal chillers was approximately 5700 tons, although at peak times was limited to 4700 tons due to cooling tower constraints. The balance of the approximately 9300-ton peak load would be served by three new 2500-ton electric centrifugal chillers, with one serving as backup. These new centrifugal chillers would replace the existing absorption chillers. The following were determined to be the advantages of Alternative 1: a new chiller enclosure was not required, new cooling towers were not required, and no TES tank (site issue). The disadvantages of Alternative 1 were as follows: a larger chiller capacity was required than other alternatives, chillers would run during peak hours resulting in expensive utility purchases, and one wall in the central plant area would need to be demolished to bring in chillers.

Alternative 2—Addition of Full Storage CHW TES

Alternative 2 analyzed the installation of a 57,000 ton-hour CHW TES tank, assumed use of the existing centrifugal chillers (4750 tons), and considered replacing the existing absorption chillers with three new 1200-ton electric centrifugal chillers. The new 1200-ton chillers would use the three existing ceramic towers for heat rejection.

The required CHW TES tank would be approximately 7.6 million gallons in volume capacity given a 12°F differential temperature (ΔT) between the CHW supply and CHW return temperatures. A 7.6 million gallon CHW TES tank could be approximately 110 ft in diameter and 110 ft in height.

With this alternative, it was recommended that the CHW system be converted to a primary-secondary configuration by adding a bypass between the CHW return and supply lines at the interface between the chillers and the distribution system. This retrofit would also require new primary CHW pumps to serve the chiller loop. The recommendation was made for further hydraulic analysis to be conducted in an effort to determine if the existing CHW pumps could be converted to variable volume secondary CHW pumps with the addition of variable frequency drives (VFDs).

The following were determined to be the advantages of Alternative 2: the chillers would run at non-peak hours resulting in high energy savings, smaller chillers were required than those in conventional chiller plants, new cooling towers were not required, and a primary-secondary pump configuration would be used. The disadvantages of Alternative 2 were listed as follows: higher excavation and construction costs, visual concern over tank location near streets, one wall in the central plant area would need to be demolished to bring in chillers.

As shown in Fig. 14-4, Table 14-5, with an approximately 57,000 ton-hour CHW TES tank, the chillers are held off during the on-peak period, and during this period the TES

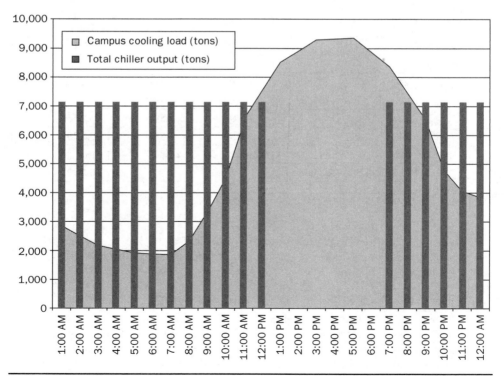

FIGURE 14-4   Alternative 2 plant operating schedule.

Hour ending	Campus load (tons)	Chiller output (tons)	TES capacity (ton-hours)
1	2,854	7,148	13,833
2	2,509	7,148	18,472
3	2,165	7,148	23,455
4	2,052	7,148	28,551
5	1,940	7,148	33,759
6	1,900	7,148	39,008
7	1,860	7,148	44,296
8	2,337	7,148	49,107
9	3,332	7,148	52,923
10	4,469	7,148	55,602
11	6,512	7,148	56,239
12	7,516	7,148	55,871
13	8,521	0	47,350
14	8,915	0	38,435
15	9,310	0	29,125
16	9,332	0	19,793
17	9,354	0	10,439
18	8,867	0	1,573
19	8,379	7,148	342
20	7,490	7,148	0
21	6,472	7,148	676
22	4,719	7,148	3,105
23	4,001	7,148	6,253
24	3,862	7,148	9,539
Total	**128,669**	**128,669**	
TES Tank Size (ton-hours)			**56,239**

TABLE **14-5** Alternative 2 Plant-Operating Schedule

tank provides the CHW supply to the campus. During the remaining portions of the day, the chillers are operated at maximum capacity to provide CHW to the campus and to charge the TES tank.

Alternative 3—Addition of Partial Storage CHW TES

Alternative 3 studied the installation of a 32,000 ton-hour CHW TES tank, assumed use of the existing centrifugal chillers (4750 tons), and considered replacing the oldest existing absorption chillers (ABS-1 and ABS-2) with two new 700-ton electric centrifugal chillers. The new 700-ton chillers would use the three existing ceramic towers for heat rejection.

The required CHW TES tank for this alternative would be approximately 4.3 million gallons in volume capacity given a 12°F differential temperature between the CHW supply and CHW return temperatures. A 4.3 million gallon CHW TES tank could be approximately 90 ft in diameter and 90 ft in height.

As with Alternative 2, it was recommended that the CHW system in this alternative be converted to a primary-secondary configuration by adding a bypass between the CHW return and supply lines at the interface between the chillers and the distribution system. This would also require new primary CHW pumps to serve the chiller loop. The recommendation was made for further hydraulic analysis to be conducted in an effort to determine if the existing CHW pumps could be converted to variable volume secondary CHW pumps with the addition of VFDs. Figure 14-5 and Table 14-6 show the partial storage TES operation.

The following were the advantages of Alternative 3: the chillers were smaller and less in number than those in conventional chiller plants, new cooling towers were not required, less space was required, and a primary-secondary pump arrangement would be required. The disadvantages of Alternative 3 were described as follows: higher excavation and construction costs than Alternative 1, visual concern over tank location near streets, and one wall in the central plant area would need to be demolished to bring in chillers.

As shown, with an approximately 32,000 ton-hour CHW TES tank, the chillers operate at an average load of approximately 5400 tons for the entire day. At times when

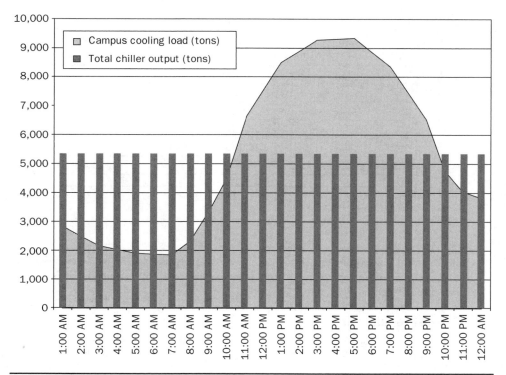

FIGURE 14-5   Alternative 3 plant-operating schedule.

Hour ending	Campus load (tons)	Chiller output (tons)	TES capacity (ton-hours)
1	2,854	5,361	6,009
2	2,509	5,361	8,861
3	2,165	5,361	12,057
4	2,052	5,361	15,366
5	1,940	5,361	18,787
6	1,900	5,361	22,248
7	1,860	5,361	25,749
8	2,337	5,361	28,773
9	3,332	5,361	30,803
10	4,469	5,361	31,695
11	6,512	5,361	30,544
12	7,516	5,361	28,389
13	8,521	5,361	25,229
14	8,915	5,361	21,675
15	9,310	5,361	17,727
16	9,332	5,361	13,756
17	9,354	5,361	9,763
18	8,867	5,361	6,258
19	8,379	5,361	3,240
20	7,490	5,361	1,111
21	6,472	5,361	0
22	4,719	5,361	642
23	4,001	5,361	2,003
24	3,862	5,361	3,502
Total	**128,669**	**128,669**	
TES tank size (ton-hours)			**31,695**

TABLE 14-6 Alternative 3 Plant-Operating Schedule

the campus load is higher than the chiller output, the deficit is drawn from the TES tank. At times when the campus load is lower than the chiller output, the surplus is put into the TES tank.

Alternative Summary

Table 14-7 summarizes each of the central chiller plant alternatives, including the operation of the Central Plant and the type and capacity of the new equipment installed.

Alternative	Additional required chillers	Chiller size (tons)	TES equipment	Other equipment
Alternative 1 Conventional chiller plant (base case)	3	2,500	N/A	3–5,000 gpm CHW Pumps
Alternative 2 Addition of full storage chilled water TES	3	1,200	57,000 ton-hour 7.6 million gal @12°F ΔT	3–2,400 gpm CHW Primary Pumps, 3–5,000 gpm Secondary Pumps
Alternative 3 Addition of partial storage chilled water TES	2	700	32,000 ton-hour 4.3 million gal @12°F ΔT	2–1,400 gpm CHW Primary Pumps, 3–5,000 gpm Secondary Pumps

TABLE 14-7   Central Chiller Plant Alternative Summary

Modeling Process

As a part of this study, each central plant alternative was modeled using a spreadsheet model. Estimated peak day heating and cooling load profiles were developed using existing campus information and used to model the peak cooling and heating load profiles. Seasonal load profiles were created and used to estimate the equipment energy consumption at various times of the day. The input data included power inputs, equipment efficiencies, and energy sources. Energy prices and TOU schedules were entered to estimate monthly and annual energy costs. The generated power from cogeneration was also included in the energy modeling. Table 14-8 provides a summary of the initial annual energy use and cost for each alternative, as determined in the spreadsheet models. .

Budget Construction Cost Estimates

A budget construction cost estimate was created for all three chiller plant alternatives evaluated in this study. The estimated budget construction costs of the alternatives included site work, concrete and masonry work, mechanical work, electrical work, and instrumentation and controls work. In addition to the direct costs associated with the above-mentioned work, the construction costs included sales tax on the cost of all materials, general requirements, 20-percent contingency, insurance and bonds, and contractor's overhead and profit.

Alternative 1—Conventional Chiller Plant (Base Case)

Table 14-9 shows the budget construction cost for Alternative 1 including the cost of the chillers, pumps, and electrical. The budget construction cost for this alternative was approximately $6 million. Most of the costs were due to the addition of three 2500-ton electric centrifugal chillers and the associated electrical equipment.

Descrip-tion	Electrical consumption (kWh)			Total consump-tion (kWh)	Electrical demand (kW)			Energy cost ($)
	On peak	Mid peak	Off peak		On peak	Mid peak	Off peak	
Alt 1	129,846	246,902	199,003	575,751	6,450	7,219	5,228	1,304,064
Alt 2	4,561	17,859	327,534	349,954	324	962	7,061	810,109
Alt 3	74,202	174,296	300,022	548,520	3,640	4,523	4,238	1,029,469

TABLE 14-8 Central Plant Energy Usage and Cost Summary

Item	Qty.	Total ($)
Electric drive centrifugal chiller, 2,500 tons	3	1,642,500
Chilled water pump, 5,000 gpm, 200 hp	3	62,400
Electrical	1	1,300,000
Budget construction cost[1]		**5,920,000**

Note 1: Includes soft cost markups and tax.

TABLE 14-9 Alternative 1 Budget Construction Cost Summary

Alternative 2—Addition of Full Storage CHW TES

Table 14-10 contains the budget construction cost for Alternative 2, including the full storage TES tank and all additional major equipment. The budget construction cost for this alternative was approximately $12.3 million, twice the amount of Alternative 1, mainly due to the full storage TES tank.

Alternative 3—Addition of Partial Storage CHW TES

Table 14-11 shows the budget construction cost for Alternative 3, including the partial storage tank and all major equipment costs. The budget construction cost for this alternative was approximately $7.5 million, roughly $1.5 million higher than Alternative 1. The budget construction cost for this alternative was lower than Alternative 2 by nearly $5 million, mainly due to the smaller required TES tank size.

Economic Analysis

There are a variety of economic factors that affect economic analysis and were used over the course of this study. These economic factors are the discount rate, the escalation rates, and system financing terms.

The discount factor is one of the most important numbers used in life-cycle-cost analysis. The number used in the discount factor equates future values with present values. In general, the discount factor should equal the long-term cost of money. For the purposes of this study, a 6 percent discount factor was used.

Item	Qty.	Total ($)
Full storage TES tank 56,000 ton-hour	1	3,586,000
Electric drive centrifugal chiller, 1,200 tons	3	810,480
Primary chilled water pump, 2,400 gpm , 100 hp	3	29,400
Secondary chilled water pump, 5,000 gpm, 200 hp	3	62,400
Electrical	1	880,000
Budget construction cost[1]		**12,340,000**

Note 1: Includes soft cost markups and tax.

TABLE 14-10   Alternative 2 Budget Construction Cost Summary

Item	Qty.	Total ($)
Partial storage TES tank 32,000 ton-hours	1	2,120,000
Electric drive centrifugal chiller, 700 tons	3	516,180
Primary chilled water pump, 2,400 gpm , 100 hp	2	19,600
Secondary chilled water pump, 5,000 gpm, 200 hp	3	62,400
Electrical	1	640,000
Budget construction cost[1]		**7,450,000**

Note 1: Includes soft cost markups and tax.

TABLE 14-11   Alternative 3 Budget Construction Cost Summary

Escalation rate factors are the rate at which various costs increase. The escalation rates used are estimates for long-term escalation (i.e., 20 years). It is important to be conservative in escalation rate estimates, so as not to overestimate electricity price increases. When this study was conducted, deregulation had made forecasting electricity escalation rates extremely challenging, as experts disagreed on the future direction of electricity prices.

For the purposes of this study, natural gas and electricity prices were assumed to escalate at 2.6 percent per year, and operating and maintenance (O&M) costs were assumed to escalate at 3.2 percent per year. The natural gas and electricity escalation rate was based on an average provided by the State Fuel Price Index for electricity and natural gas. The operation and maintenance escalation rate was based on the consumer price index for all urban consumers.

Life-Cycle Costs

Life-cycle-cost analysis is a process by which system costs are calculated not just for a particular period, but for the life of the system. In addition, life-cycle-cost analysis is a process by which the time value of money is taken into consideration. The life cycle used for this study is a 20-year period.

Description	Project cost ($)	First- year maintenance costs ($)	First year energy cost ($)	First year annual costs ($)	20 year NPV ($)	NPV savings ($)
Alternative 1	5,920,000	132,758	1,304,064	1,438,823	26,284,780	—
Alternative 2	12,340,000	126,850	810,109	938,959	25,656,224	628,557
Alternative 3	7,450,000	117,479	1,029,469	1,148,948	23,719,239	2,565,541

TABLE 14-12   Life-Cycle-Cost Analysis Summary

The life-cycle-cost estimating process used in this study involved estimating the annual costs and savings that would result from each of the evaluated alternatives for each year in the life-cycle period. Costs and savings change from 1 year to the next due to escalation of various parameters such as electric and natural gas purchase costs. Given each year's costs and savings, an annual net cash flow was calculated. It was assumed that there is no equipment salvage value at the end of year 20. Given a cash flow and a discount factor, the NPV was then calculated.

Net Present Value Calculations

The use of net present value (NPV) calculations accounts for the time value of money. The premise is that, due to inflation and interest earnings, money received in the future is worth less than money received now. This study calculated the present value of each future year cash flow expense resulting from each alternative and then added each of these present values together to obtain a single NPV, providing a single number with which to compare the various cases on an equal basis.

Life-Cycle-Cost Summary

Table 14-12 provides the capital cost, first-year energy cost, first-year O&M cost, and the 20-year NPV for the three chiller plant alternatives analyzed in this study. The NPV ranges from approximately $23 to $26 million. It is important to note that the best case, Alternative 3, which has approximately $2.6 million of NPV savings over Alternative 1, has the lowest NPV. The partial storage TES alternative has the lowest NPV due in part to smaller chillers and storage tank, which has a smaller capital cost.

Conclusions and Recommendations

The focus of this study was to evaluate the central plant and TES alternatives to determine the most cost-effective cooling system option for the university. The life-cycle-cost analysis indicated that the TES alternatives were the most cost-effective options.

This Chiller Plant Expansion or TES study confirmed that partial storage TES, Alternative 3, was the most economical alternative with an NPV of approximately $23.7 million. It was determined that Alternative 3 would provide sufficient cooling capacity to meet the projected cooling demands and significant energy cost savings for the university, and was, therefore, the recommended alternative.

In addition to recommending Alternative 3, other key recommendations for the university to implement in the central plant were the following:

- Convert from constant volume primary only to variable volume system.
- Add a modern central plant DDC control system.
- Conduct CHW system hydronic analysis to verify adequacy of distribution piping.
- Increase CHW delta-T to increase capacity and save energy (e.g., convert three-way valves to two-way valves).

CHAPTER 15

Ice Thermal Storage Case Study

Study Objective

A thermal energy storage (TES) study was commissioned by a large entertainment facility to determine the technical and economic feasibility of installing and operating an ice TES system to serve the facility's cooling loads. The facility was planning a major expansion and, as such, the facility's cooling loads were expected to increase by approximately 1000 tons. The facility was served by an existing central chiller plant, which provided chilled water (CHW) at approximately 43°F. The CHW system typically operated at about 12°F delta-T (55°F CHWR temperature). The facility was considering adding an additional electric-drive centrifugal chiller to account for the expected increased cooling loads and wanted to review the option of adding an ice TES system, which would eliminate the need for the additional chiller, provide for on-peak electric power demand reduction and energy cost savings, and would also provide significant rebates from the local utility company.

Central Plant Facilities

The existing facility's central plant was constructed in the late 1990s. The central plant housed the central cooling systems that served buildings and cooling loads at the facility.

Chiller Plant

There were three existing 1300-ton nominal electric centrifugal chillers located in the existing central plant, when this study was conducted. One of these chillers served as a backup in case one of the other chillers was out of commission. All three chillers appeared to be in good condition; however, there were some staff reports that the chillers were unable to meet the reported capacity. The chillers were selected to generate CHW at approximately 45° F with a design delta-T of approximately 16°F at design conditions but it had been noted by the facility's staff that the chillers were set for a CHW supply temperature of 43ºF and were operated at a 12°F delta-T. The existing chillers are described in detail in Table 15-1.

Tag	CH-1	CH-2	CH-3
Equipment type	Electric centrifugal	Electric centrifugal	Electric centrifugal
Date of manufacture	1996	1996	1996
Nominal cooling capacity (tons)	1,300	1,300	1,300
Kw/ton (at design conditions)	0.544	0.544	0.544
Chilled water flow rate (gpm)	1,950	1,950	1,950
Chilled water EWT (°F)	61	61	61
Chilled water LWT (°F)	45	45	45
Condenser water flow rate (gpm)	3,900	3,900	3,900
Condenser water EWT (°F)	85	85	85
Condenser water LWT (°F)	94.3	94.3	94.3
Refrigerant type	R-134a	R-134a	R-134a
Input kW	707	707	707
Electrical (V-Φ-Hz)	4160-3-60	4160-3-60	4160-3-60

TABLE 15-1   Chiller Summary

Chilled Water Pumping

The existing facility's CHW system was a constant-volume primary with variable-volume secondary pumping system. A simple schematic of the CHW pumping system is shown in Fig. 15-1. The pumping system consisted of primary CHW pumps that operated at constant speeds and pumped constant volumes of water individually through each chiller, and secondary CHW pumps that operated with variable-frequency drives (VFDs) to modulate speeds and to vary pump flows and pressures. The VFDs modulated the secondary CHW pumps to maintain differential pressure across the most hydraulically remote CHW load.

Central Plant CHW Pumps

The facility installed three primary CHW pumps (PP) to serve the centrifugal chillers located in the existing central plant. The CHW pumping system within the existing facility's central plant included three constant-volume primary CHW pumps (PP-1, 2 & 3) and three variable-volume secondary CHW pumps (SP-1, 2 & 3). Each primary CHW pump served its respective chiller and the pumps were not interconnected together with a common header. Each primary pump was double-suction split-case type, rated for 1950 gpm and 50 ft of total dynamic head (TDH) using a 40-hp motor. The variable-volume secondary CHW pumps were connected together with a common header to provide CHW to the facility's buildings and loads as required. Each secondary pump was double-suction split-case type, rated for 1950 gpm and 210 ft TDH using a 200 hp inverter duty-type motor.

Central Plant Cooling Towers

Three existing metal-framed cooling towers provided cooling water to the condensers of centrifugal chillers in the existing central plant. The cooling towers were located on

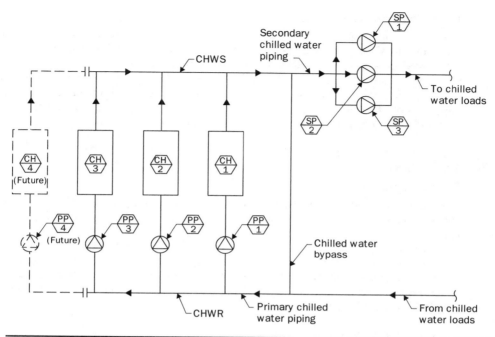

FIGURE 15-1   Constant-volume primary with variable-volume secondary CHW pumping system schematic for the facility.

grade and were rated at a flow of 3900 gpm each. The towers were designed to operate at condenser water inlet and outlet temperatures of approximately 95°F and 85°F, respectively. The design peak wet bulb temperature for the cooling towers was 75°F. The cooling towers were induced draft cross-flow type, each with a 75-hp fan motor.

Central Plant Cooling Tower Pumps
The condenser side of the chillers and the cooling towers were connected through condenser water piping, served by the condenser water pumps. Three vertical split case cooling tower pumps served the three cooling towers. All three cooling tower pumps were located in the existing central plant's cooling tower yard adjacent to the central plant building. Each cooling tower pump was rated for 3900 gpm at 80 ft. TDH with a 125-hp motor at 1800 rpm. Each pump served one of the cooling towers.

Chilled Water Distribution System
The CHW distribution system was approximately 10 years old and in good condition at the time of this study. CHW was supplied to the facility through a 20-in. main from the existing facility's central plant. Most of the CHW network was direct buried. The CHW piping, ranging in size from 8 to 18 in., branched out at various locations within the piping network. CHW was delivered to the designated buildings through direct piping connections to air-handling and fan coil units stationed throughout the buildings and the site.

Plant Controls or Metering

The plant control system had the ability to start and stop central plant equipment from workstations within the plant. Plant operators handled control of equipment in terms of scheduling and sequencing with an energy management system. The facility had extensive metering capabilities, including flows and temperatures into and out of the central plant and most buildings onsite. Each served building typically included CHW supply and CHW return temperature sensors, as well as flow meters. These meters were all used to measure individual building CHW energy consumption.

Cooling Loads

Existing Cooling Loads

The facility provided the recorded existing peak day cooling load profiles for use in this study. Table 15-2 presents a summary of the provided peak cooling loads for the existing conditions at the facility.

As shown in Table 15-2, the existing peak design day CHW load at the facility was approximately 2100 tons and it occurred at 3:00 PM, which was coincident with the local utility company's on-peak electrical demand period that occurred between 1:00 and 5:00 PM each summer day. The existing minimum design day CHW load at the facility was 694 tons and it occurred at 2:00 AM.

Existing chilled water cooling loads	
Time of day	**Design day cooling loads (tons)**
12:00 AM	789
1:00 AM	738
2:00 AM	694
3:00 AM	706
4:00 AM	702
5:00 AM	701
6:00 AM	923
7:00 AM	983
8:00 AM	1,055
9:00 AM	1,284
10:00 AM	1,576
11:00 AM	1,815
12:00 PM	1,932
1:00 PM	2,053
2:00 PM	2,097

TABLE 15-2   Existing Reported Cooling Loads for the Facility

Existing chilled water cooling loads	
Time of day	**Design day cooling loads (tons)**
3:00 PM	2,100
4:00 PM	2,016
5:00 PM	1,874
6:00 PM	1,700
7:00 PM	1,504
8:00 PM	1,409
9:00 PM	1,330
10:00 PM	1,239
11:00 PM	947
Total	**32,167**

TABLE 15-2   Existing Reported Cooling Loads for the Facility (*continued*)

Future Cooling Loads

The existing design day CHW load profile was used to estimate the future peak CHW load and coincident future peak day hourly load profile. Since the existing design day CHW loads were based upon observed conditions, it was determined that the existing CHW loads were inherently diversified.

The future peak day cooling load and coincident load profile for the planned expansion were estimated using information provided by the facility's management. The facility's management had stated that the planned expansion was expected to add 1000 tons (non-diversified) of cooling load. An 80-percent diversity factor was applied to the 1000-ton peak cooling load increase, yielding an adjusted peak cooling load increase seen at the facility's central plant of 800 tons. Table 15-3 provides a summary of the future cooling load profile foreseen at the facility.

As shown in Table 15-3, for each hour of the existing facility's cooling load profile, the existing cooling load was divided by the existing peak cooling load and multiplied by 100. The resulting values represent each hour's percentage of the peak cooling load.

The adjusted additional cooling load profile for the extreme day at the facility was developed by assuming that the adjusted peak additional cooling load of 800 tons would occur coincident with the existing peak cooling load that occurred at 2:00 and 3:00 PM on the existing peak day. The remaining adjusted additional cooling loads for each hour of the day were then calculated by multiplying the adjusted peak additional cooling load by the percent of existing peak cooling load for the corresponding hour.

The estimated future cooling load profile was then developed by adding the adjusted additional cooling load to the existing cooling load for each hour of the day. As shown in Table 15-3, the estimated peak future cooling load for the facility was approximately 2900 tons and occurred at 3:00 PM, which was coincident with the local utility company's on-peak electrical demand period of 1:00 to 5:00 PM.

The estimated future cooling load profile was used as the basis for the TES calculations.

Time of day	Existing cooling loads (tons)	Percent of existing load for each hour (%)	Adjusted additional cooling loads (tons)	Estimated future cooling loads (tons)
12:00 AM	789	37.6	300	1,089
1:00 AM	738	35.1	280	1,018
2:00 AM	694	33.0	264	958
3:00 AM	706	33.6	269	975
4:00 AM	702	33.4	266	968
5:00 AM	701	33.4	266	967
6:00 AM	923	44.0	351	1,274
7:00 AM	983	46.8	374	1,357
8:00 AM	1,055	50.2	401	1,456
9:00 AM	1,284	61.1	489	1,773
10:00 AM	1,576	75.0	600	2,176
11:00 AM	1,815	86.4	691	2,506
12:00 PM	1,932	92.0	735	2,667
1:00 PM	2,053	97.8	782	2,835
2:00 PM	2,097	99.9	798	2,895
3:00 PM	2,100	100.0	800	2,900
4:00 PM	2,016	96.0	768	2,784
5:00 PM	1,874	89.2	714	2,588
6:00 PM	1,700	81.0	648	2,348
7:00 PM	1,504	71.6	573	2,077
8:00 PM	1,409	67.1	536	1,945
9:00 PM	1,330	63.3	506	1,836
10:00 PM	1,239	59.0	471	1,710
11:00 PM	947	45.1	360	1,307
Total	**32,167**		**12,242**	**44,409**

TABLE 15-3   Estimated Future Cooling Loads for the Facility

Implementing Ice TES at the Facility

Originally constructed in the 1990s, the existing facility's central plant included three electric-driven centrifugal chillers that were designed to chill water to 43°F. To increase the facility's existing CHW plant capacity and serve future CHW loads, a study was commissioned by facility's management to evaluate implementing the following alternatives:

- Alternative 1—Base Case
- Alternative 2—Full Storage 4-Hours Ice TES system

- Alternative 3—Full Storage 12-Hours Ice TES System
- Alternative 4—Partial Storage Ice TES System
- Alternative 5—Full Storage Ice Ball TES System

At the onset of this study, the facility's management provided several options to evaluate as part of the TES system alternatives considered in this study. One option used a large water feature, located in the center of the facility's property, as the source of condenser water for the proposed ice TES system with a corresponding water temperature of approximately 78°F; another option considered the use of buried concrete ice TES tanks.

Alternatives

Alternative 1—Base Case

This alternative analyzed the energy cost of operating a conventional central chiller plant at the projected future load. Since the future peak cooling load exceeded the capacity of the facility's central plant, one additional 1300-ton chiller was proposed to be added to the existing central chiller plant in this alternative. Additional ancillary equipment and increased CHW header sizes were also considered in Alternative 1.

Alternative 2—Full Storage 4-Hours Ice TES system

Alternative 2 analyzed a full storage, external melt, ice TES system, building ice over 20 hours for use during the four on-peak hours of each day. The central ice TES system would include the three existing 1300-ton electric centrifugal chillers, existing primary pumps, existing secondary pumps and a new 7000-ton-hour ice TES system. The ice TES system would consist of two underground ice storage tanks with ice coils (3500 ton-hours each tank), one 350-ton water-cooled ice-making chiller, one plate and frame water-to-water heat exchanger, pumps, and appurtenances.

Figure 15-2 presents a simple CHW piping schematic for Alternative 2. The schematic shows the ice TES system equipment connected to the existing facility's CHW piping via 16-in. diameter CHW piping. Note that required isolation valves, balancing valves, and controls are not shown for clarity and simplicity.

With the ice TES system in "charge" mode, the ice TES tanks would be charged by the ice chiller, and the CHW pumps and the CHW/ice tank pumps would not operate and CHW would not be circulated through the ice TES tanks. With the TES system in "discharge" mode, the ice chiller would not operate and the CHW/ice tank pumps would circulate water through the ice tanks to melt the ice and generate cold CHW for delivery to the water-to-water heat exchanger. The CHW pumps would pump CHW return to the water-to-water heat exchanger into the existing facility's CHW supply piping loop. The water-to-water heat exchanger isolated the ice TES system from the existing facility's CHW piping loop.

The approximate available ice TES capacity was calculated based upon the given ice-making chiller's maximum nominal capacity and the number of charging mode operating hours occurring each day. Using these values, the approximate available ice TES capacity was calculated to be approximately 7000 ton-hours. The full storage ice TES operational strategy for Alternative 2 on a peak day is shown in Table 15-4.

As shown in Table 15-4, during the 4 hours of the on-peak electric demand period, the ice-making chiller would not operate. The ice TES system would be ultimately capable of delivering 2675 tons of CHW during its first hour of operation, 1445 tons of

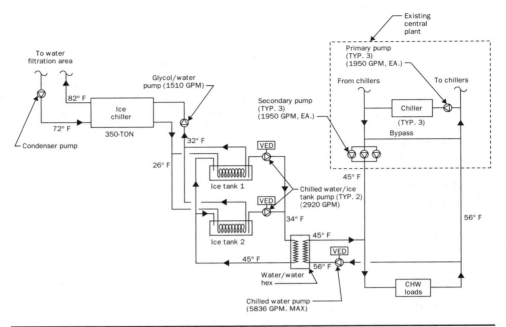

FIGURE 15-2 Ice thermal storage piping schematic.

Alternative 2—Full Storage 4-Hour Ice TES Storage					
Hour	Facility's chilled water cooling load (tons)	Centrifugal chiller output (tons)	Ice-making chiller output (tons)	Ice TES capacity (ton-hours)	Mode
1:00 AM	963	963	350	3,150	Charge
2:00 AM	906	906	350	3,500	Charge
3:00 AM	921	921	350	3,850	Charge
4:00 AM	916	916	350	4,200	Charge
5:00 AM	915	915	350	4,550	Charge
6:00 AM	1,204	1,204	350	4,900	Charge
7:00 AM	1,283	1,283	350	5,250	Charge
8:00 AM	1,377	1,377	350	5,600	Charge
9:00 AM	1,675	1,675	350	5,950	Charge
10:00 AM	2,056	2,056	350	6,300	Charge
11:00 AM	2,368	2,368	350	6,650	Charge
12:00 PM	2,521	2,521	350	7,000	Charge
1:00 PM	2,675	0	0	4,325	Discharge
2:00 PM	2,740	1,295	0	2,880	Discharge

TABLE 15-4 Peak Day Thermal Storage Calculations for Alternative 2—Full Storage 4-Hours Ice

Alternative 2—Full Storage 4-Hour Ice TES Storage				
Hour	Facility's chilled water cooling load (tons)	Centrifugal chiller output (tons)	Ice-making chiller output (tons)	Ice TES capacity (ton-hours)	Mode
3:00 PM	2,740	1,300	0	1,440	Discharge
4:00 PM	2,630	1,190	0	0	Discharge
5:00 PM	2,445	2,445	350	350	Charge
6:00 PM	2,218	2,218	350	700	Charge
7:00 PM	1,962	1,962	350	1,050	Charge
8:00 PM	1,838	1,838	350	1,400	Charge
9:00 PM	1,735	1,735	350	1,750	Charge
10:00 PM	1,617	1,617	350	2,100	Charge
11:00 PM	1,236	1,236	350	2,450	Charge
12:00 AM	1,029	1,029	350	2,800	Charge
Total (ton-hours)	41,970	34,970	7,000		
Required TES capacity (ton-hours)				7,000	
Required chiller capacity (tons)		2,521	350		

TABLE 15-4   Peak Day Thermal Storage Calculations for Alternative 2—Full Storage 4-Hours Ice (*continued*)

CHW during its second hour of operation, and 1440 tons of CHW during its third and fourth hours of operation for a total of 7000 ton-hours. The ice TES tanks would be completely discharged at the end of the on-peak electric demand period and the ice-making chiller would commence operating to make ice as the mid-peak electric demand period began to restart the cycle.

As shown in Table 15-4, the existing chillers in the existing central plant would not operate during the first hour of the on-peak electric demand period. For the remaining three hours of the on-peak electric demand period, only one existing chiller (1300 tons, nominal) in the existing central plant would operate in conjunction with the remaining ice stored in the ice TES tanks to serve the facility's CHW cooling load demand.

Alternative 3—Full Storage 12-Hours Ice TES system
Alternative 3 analyzed a full storage ice TES system, building ice during 12 mid-peak and off-peak hours (from 10:00 PM to 11:00 AM), for use during the four on-peak hours of each day. The general system arrangement for Alternative 3 is similar to that of Alternative 2. The ice-making chiller output is estimated at 584 tons. The hourly ice storage capacity, in ton-hours, is shown in Table 15-5.

Alternative 4—Partial Storage Ice TES system
Alternative 4 analyzed a partial storage ice TES system, building ice during all hours of the day (on-peak as well as mid-peak and off-peak). The general system arrangement for Alternative 4 is similar to that of Alternative 2 and 3. The ice-making chiller output is estimated at 292 tons. The hourly ice storage capacity, in ton-hours, is shown in Table 15-6.

Alternative 3—Full Storage 12-Hour Ice TES Storage					
Hour	Facility's chilled water cooling load (tons)	Centrifugal chiller output (tons)	Ice-making chiller output (tons)	Ice TES capacity (ton-hours)	Mode
1:00 AM	963	963	584	2,336	Charge
2:00 AM	906	906	583	2,919	Charge
3:00 AM	921	921	583	3,502	Charge
4:00 AM	916	916	583	4,085	Charge
5:00 AM	915	915	583	4,668	Charge
6:00 AM	1,204	1,204	583	5,251	Charge
7:00 AM	1,283	1,283	583	5,834	Charge
8:00 AM	1,377	1,377	583	6,417	Charge
9:00 AM	1,675	1,675	583	7,000	Charge
10:00 AM	2,056	2,056	0	7,000	Standby
11:00 AM	2,368	2,368	0	7,000	Standby
12:00 PM	2,521	2,521	0	7,000	Standby
1:00 PM	2,675	0	0	4,325	Discharge
2:00 PM	2,740	1,295	0	2,880	Discharge
3:00 PM	2,740	1,300	0	1,440	Discharge
4:00 PM	2,630	1,190	0	0	Discharge
5:00 PM	2,445	2,445	0	0	Standby
6:00 PM	2,218	2,218	0	0	Standby
7:00 PM	1,962	1,962	0	0	Standby
8:00 PM	1,838	1,838	0	0	Standby
9:00 PM	1,735	1,735	0	0	Standby
10:00 PM	1,617	1,617	584	584	Charge
11:00 PM	1,236	1,236	584	1,168	Charge
12:00 AM	1,029	1,029	584	1,752	Charge
Total (ton-hours)	41,970	34,970	7,000		
Required TES capacity (ton-hours)				7,000	
Required chiller capacity (tons)	2,521	584			

TABLE 15-5 Peak Day Thermal Storage Calculations for Alternative 3—Full Storage 12-Hours Ice

Alternative 4—Partial Storage Ice TES Storage					
Hour ending	Facility's chilled water cooling load (tons)	Centrifugal chiller output (tons)	Ice-making chiller output (tons)	Ice TES capacity (ton-hours)	Mode
1:00 AM	963	963	292	2,625	Charge
2:00 AM	906	906	292	2,917	Charge
3:00 AM	921	921	292	3,209	Charge
4:00 AM	916	916	292	3,501	Charge
5:00 AM	915	915	292	3,793	Charge
6:00 AM	1,204	1,204	292	4,085	Charge
7:00 AM	1,283	1,283	292	4,377	Charge
8:00 AM	1,377	1,377	292	4,669	Charge
9:00 AM	1,675	1,675	292	4,961	Charge
10:00 AM	2,056	2,056	292	5,253	Charge
11:00 AM	2,368	2,368	292	5,545	Charge
12:00 PM	2,521	2,521	291	5,836	Charge
1:00 PM	2,675	0	291	3,452	Discharge
2:00 PM	2,740	1,295	291	2,298	Discharge
3:00 PM	2,740	1,300	291	1,149	Discharge
4:00 PM	2,630	1,190	291	0	Discharge
5:00 PM	2,445	2,445	291	291	Charge
6:00 PM	2,218	2,218	291	582	Charge
7:00 PM	1,962	1,962	291	873	Charge
8:00 PM	1,838	1,838	292	1,165	Charge
9:00 PM	1,735	1,735	292	1,457	Charge
10:00 PM	1,617	1,617	292	1,749	Charge
11:00 PM	1,236	1,236	292	2,041	Charge
12:00 AM	1,029	1,029	292	2,333	Charge
Total (ton-hours)	41,970	34,970	7,000		
Required TES capacity (ton-hours)				5,836	
Required chiller capacity (tons)		2,521	292		

TABLE 15-6   Peak Day Thermal Storage Calculations for Alternative 4—Partial Storage Ice

Alternative 5—Full Storage Ice Ball TES system

Alternative 5 analyzed a full storage ice ball TES system, building ice over 20 hours, for use during the four on-peak hours of each day. This alternative is similar to Alternative 2 in nature, utilizing a similar operational strategy; however, Alternative 5 uses an ice ball TES system as an alternative to the ice-harvesting TES system of Alternative 2. The ice-making chiller capacity is estimated at 350 tons. The ice storage capacity, in ton-hours, is shown in Table 15-7.

Alternative 5—Full Storage 4 -Hour Ice Ball TES Storage					
Hour ending	**Facility's chilled water cooling load (tons)**	**Centrifugal chiller output (tons)**	**Ice-making chiller output (tons)**	**Ice TES capacity (ton-hours)**	**Mode**
1:00 AM	963	963	350	3,150	Charge
2:00 AM	906	906	350	3,500	Charge
3:00 AM	921	921	350	3,850	Charge
4:00 AM	916	916	350	4,200	Charge
5:00 AM	915	915	350	4,550	Charge
6:00 AM	1,204	1,204	350	4,900	Charge
7:00 AM	1,283	1,283	350	5,250	Charge
8:00 AM	1,377	1,377	350	5,600	Charge
9:00 AM	1,675	1,675	350	5,950	Charge
10:00 AM	2,056	2,056	350	6,300	Charge
11:00 AM	2,368	2,368	350	6,650	Charge
12:00 PM	2,521	2,521	350	7,000	Charge
1:00 PM	2,675	0	0	4,325	Discharge
2:00 PM	2,740	1,295	0	2,880	Discharge
3:00 PM	2,740	1,300	0	1,440	Discharge
4:00 PM	2,630	1,190	0	0	Discharge
5:00 PM	2,445	2,445	350	350	Charge
6:00 PM	2,218	2,218	350	700	Charge
7:00 PM	1,962	1,962	350	1,050	Charge
8:00 PM	1,838	1,838	350	1,400	Charge
9:00 PM	1,735	1,735	350	1,750	Charge
10:00 PM	1,617	1,617	350	2,100	Charge
11:00 PM	1,236	1,236	350	2,450	Charge
12:00 AM	1,029	1,029	350	2,800	Charge
Total (ton-hours)	**41,970**	**34,970**	**7,000**		
Required TES capacity (ton-hours)				**7,000**	
Required chiller capacity (tons)		**2,521**	**350**		

TABLE 15-7   Peak Day Thermal Storage Calculations for Alternative 5—Full Storage Ice Ball

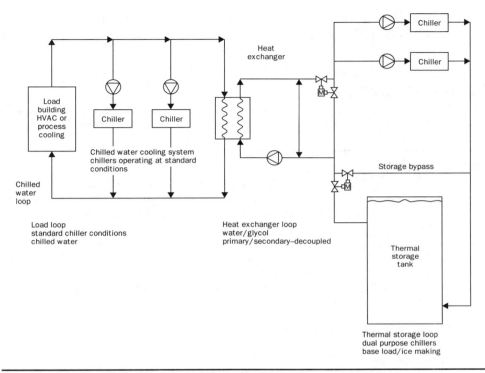

FIGURE 15-3   Ice thermal storage system schematic.

A general schematic of the central ice TES system for Alternative 5 is shown in Fig. 15-3.

Ice TES System Site

The ice TES system was proposed to include one 7000-ton-hour, buried concrete ice TES tank complete with ice-making coils, one 350-ton ice-making chiller (based on Alternative 2), one water-to-water plate and frame heat exchanger, and all required pumps. Facility staff suggested locating the ice TES system within the facility, buried near a guests' attraction, along with the plate and frame heat exchanger and two ice TES tank CHW pumps. Additional equipment, including the ice-making chiller and pumps, was proposed to be installed outside of the facility's boundary, near support facilities.

Noted advantages to locating the ice TES system equipment in the facility location included: buried ice TES tank and designated equipment would not be an aesthetic nuisance to facility's guests; power required for the ice-making chiller and pumps was readily available due to the proximity of the proposed ice TES equipment site to the electrical service; and the ice-making chiller and designated pumps would be hidden from facility view by an existing 16-ft tall concrete block wall.

Along with the advantages to locating the ice TES system equipment in this area, there were noted disadvantages that included: constructing a buried ice TES tank near an existing attraction would prove to be incredibly challenging and expensive and the distance between the proposed ice TES chiller location and the proposed condenser water connection to the large water feature would require long lengths of condenser water piping.

Utility Rate Structure

The facility operated on a time of use (TOU) rate schedule. Table 15-8 summarizes the local utility company's TOU rate schedule at the time of this study.

Energy Analysis

The annual electricity consumption and electricity cost was calculated and analyzed for Alternatives 1 and 2. Energy Analysis for Alternatives 3, 4, and 5 was deemed unnecessary by the facility's management since these alternatives were not cost-effective for the facility.

The estimated future peak daily CHW load profile for the facility was used to model the summer peak energy demand and estimate the on-peak and mid-peak electricity demand in kilowatts (kW) for Alternative 1 and Alternative 2. The on-peak and mid-peak electricity demands were used to calculate the summer on-peak and mid-peak demand charges, respectively.

Energy Consumption and Cost

Alternative 1—Base Case

As shown in Table 15-9, the annual energy consumption associated with operating Alternative 1 was approximately 8.04 million kWh.

Based upon the utility costs discussed above and the stated annual energy uses, Table 15-10 shows the annual cost for the utilities associated with operating Alternative 1, which was approximately $795,000.

Alternative 2—Full Storage 4-Hours Ice TES System

As shown in Table 15-11, the annual energy consumption associated with operating Alternative 2 was approximately 7.383 million kWh, approximately 600,000 kWh less than Alternative 1.

	Non-Time-related demand charge ($/kW)	Summer		Winter	
		Demand charge ($/kW)	Energy charge ($/kW)	Demand charge ($/kW)	Energy charge ($/kW)
On-Peak	—	10.97	0.1008	—	—
Mid-Peak	2.34	3.99	0.0761	4.20	0.0823
Off-Peak	—	—	0.0453	—	0.0453

TABLE 15-8 TOU Electricity Rate Schedule

Season	Weekday energy consumption (kWh)			Weekend energy consumption (kWh)	Energy demand (kW)	
	On-Peak	Mid-Peak	Off-Peak	Off-Peak	On-Peak	Mid-Peak
Summer peak	—	—	—	—	2,200	2,000
Summer	179,000	359,000	163,000	337,000	2,100	1,900
Summer shoulder	133,000	258,000	108,000	240,000	1,600	1,400
Winter	21,000	50,000	40,000	53,000	—	200
Winter shoulder	133,000	258,000	108,000	240,000	—	1,600
Annual totals	**1,398,000**	**2,775,000**	**1,257,000**	**2,610,000**	—	—
	8,040,000					

TABLE 15-9   Monthly Energy Consumption for the Base Case

	Summer peak	Summer ($)	Shoulder—summer rates ($)	Winter ($)	Shoulder—winter rates ($)
Weekday electric energy cost	—	59,900	43,100	8,400	39,300
Weekend electric energy cost	—	17,300	12,300	2,700	12,300
On-Peak demand charge	27,800	26,600	19,300	-	-
Mid-Peak demand charge	8,900	8,500	6,500	1,200	7,400
Non-Time-related demand charge	5,900	5,700	4,100	600	4,100
Monthly total	**42,600**	**118,000**	**85,300**	**12,900**	**63,100**
Annual total	**795,000**				

TABLE 15-10   Monthly Energy Costs for Alternative 1

Season	Energy consumption (kWh)			Weekend Energy consumption (kWh)	Energy demand (kW)	
	On-Peak	Mid-Peak	Off-Peak	Off-Peak	On-Peak	Mid-Peak
Summer peak	—	—	—	—	1,300	2,300
Summer	79,000	366,000	211,000	315,000	1,100	1,900
Summer shoulder	34,000	265,000	156,000	218,000	700	1,500
Winter	2,000	9,000	86,000	47,000	—	200
Winter shoulder	34,000	265,000	156,000	218,000	—	1,500
Annual totals	**447,000**	**2,715,000**	**1,827,000**	**2,394,000**	—	—
	7,383,000					

TABLE 15-11   Monthly Energy Consumption for Alternative 2

Based on the utility costs discussed above and the stated annual energy use, Table 15-12 shows the annual cost for the utilities associated with operating Alternative 2, which was determined to be approximately $655,000.

Budget Construction Costs

The estimated budget construction cost estimates included site work, tank construction, mechanical equipment, trenching, piping, controls, and electrical work. In addition to the direct costs associated with the above-mentioned work, the construction costs included 8.75-percent local sales tax on the cost of all materials, 5-percent general requirements on the total direct construction cost, 20-percent contingency on the total direct construction cost, 3 percent of construction cost for insurance and bonds, and 15-percent of construction cost for contractor's overhead and profit. The budget construction costs did not include project costs to account for engineering, owner contract administration during construction, inspection services, laboratory testing, and start-up or commissioning. Detailed budget construction cost summaries were generated for adding a fourth 1300-ton chiller to the existing facility's central plant (Alternative 1—Base Case) and adding an ice TES system complete with all necessary ice-making equipment and associated piping (Alternative 2—Full Storage 4-Hours Ice TES system). The budget construction cost for Alternative 1 was estimated at approximately $1.69 million, while for Alternative 2 it was estimated at approximately $4.91 million. Table 15-13 shows the budget construction cost summary of Alternative 2, which involved adding a new 7000-ton-hour underground ice TES tank, ice chiller, plate and frame heat exchanger, and associated piping.

It was determined that if the Full Storage 4-Hours ice TES alternative (Alternative 2) was to proceed, soil improvement requirements, chemical treatment costs, and electrical service costs would need further investigation to establish a more exact budget construction cost estimate.

	Summer peak ($)	Summer ($)	Shoulder— summer rates ($)	Winter ($)	Shoulder—winter rates ($)
Weekday electric energy cost	—	51,500	34,700	5,500	33,700
Weekend electric energy cost	—	16,200	11,200	2,400	11,200
On-Peak demand charge	16,400	14,100	8,500	—	—
Mid-Peak demand charge	10,400	8,600	6,700	1,000	7,000
Non-Time-related demand charge	3,500	3,000	1,800	500	3,900
Monthly total	**30,300**	**93,400**	**62,900**	**9,400**	**55,800**
Annual total	**655,000**	—	—	—	—

TABLE 15-12 Monthly Energy Cost for Alternative 2

Item	Alternative 1 cost ($)	Alternative 2 cost ($)
Site work	350,000	350,000
Ice TES tank	—	1,310,000
Additional ice equipment	—	780,000
Pumps	80,000	150,000
Piping	550,000	550,000
Controls	50,000	50,000
Electrical	80,000	80,000
Alternative 2 estimate	1,110,000	3,270,000
General requirements (5%)	55,500	163,500
Subtotal	**1,170,000**	**3,440,000**
Contingency (20%)	234,000	688,000
Subtotal	**1,410,000**	**4,130,000**
Insurance and bonds (3%)	42,300	123,900
Subtotal	**1,460,000**	**4,260,000**
Contractor's overhead and profit (15%)	221,920	647,520
Budget construction cost estimate	**$1,690,000**	**$4,910,000**

TABLE 15-13   Alternative 2 Budget Construction Cost Summary

Description	Project cost ($)	Utility incentive ($)	First-Year electricity cost ($)	Electricity cost savings ($)	Annual cost savings ($)	20-Year NPV ($)
Alternative 1	1,690,000	0	795,000	—	—	12,449,000
Alternative 2	4,910,000	1,800,000	655,000	140,000	142,000	13,337,000

TABLE 15-14   Life-Cycle-Cost Analysis Summary for Two Alternatives

Life-Cycle-Cost Summary

Table 15-14 provides the capital cost, first-year energy cost, and the 20-year net present value (NPV) for Alternative 1 and Alternative 2. The NPVs were approximately $12.449 million for the Base Case and $13.337 million for Alternative 2.

Conclusion

Alternative 2 would shift approximately 900 kW of power demand from on-peak to mid-peak and off-peak periods, as determined by subtracting the summer on-peak energy demand of Alternative 2 from the summer on-peak energy demand of Alternative 1. The 900 kW shift would allow for reduced central plant usage from 1:00 to 5:00 PM during the summer months.

Alternative 2 would also save about 657,000 kWh of energy annually when compared to Alternative 1. The energy savings would be realized through more efficient operation of the refrigeration equipment at night when the ambient dry bulb and wet bulb temperatures were lower than that during the day time, and also through the fact that chiller plant equipment would operate more frequently at full load with higher efficiencies and less parasitic load when compared to Alternative 1.

The total annual operating cost of Alternative 2 was about $140,000 less than that of Alternative 1. The initial cost of Alternative 1 was approximately $1.69 million, whereas the initial cost of Alternative 2 was approximately $4.91 million, resulting in the initial cost difference of approximately $3.22 million.

The local utility company offered an incentive program for an off-peak cooling system with a rebate of $2000 per kilowatt shifted from on-peak to off-peak periods, resulting in a total of $1,800,000 in rebates for the large entertainment facility. This rebate would be paid in the form of three annual payments starting at the completion of the TES system commissioning. The simple payback of Alternative 2 was calculated by subtracting the project cost of Alternative 1 and the utility rebate from the project cost of Alternative 2 and dividing the total by the yearly savings. The simple payback of Alternative 2 was calculated to be about 10 years.

Glossary

The following are some of the key terms one needs to understand when working with thermal storage facilities.

Active thermal storage   Active thermal storage systems use mechanical energy or mechanical transport heat transfer fluids to achieve thermal storage. One example of an active thermal storage system is a chilled water TES tank.

British thermal unit (Btu)   One Btu is the amount of heat required to raise one pound mass of water 1°F.

Charge   Charge occurs in a thermal storage tank when the plant output is greater than the load being served. The excess plant output is stored in the thermal storage tank for later discharge. The tank charge also refers to the amount of available capacity, as in it the tank is fully charged or only half charged, for example.

CHP   Combined heat and power (CHP), also known as *cogeneration*, is the simultaneous production of heat and the generation of power (typically electric power) from a single fuel source. Operating a car heater on a cold day is a form of CHP.

Coefficient of performance (COP)   The coefficient of performance is the ratio of heat transferred to the work required, in consistent units.

Conduction   Conduction is the form of heat transfer that occurs through a material(s) as atomic vibrations are transmitted through the material. One example of conductive heat transfer is the heat felt from touching a hot cooking pan.

Convection   Convection is the form of heat transfer that occurs due to the mixing and movement of fluids (gases and liquids). Natural convection allows for currents to be created by the movement of fluids with different temperatures and corresponding different densities. One example of natural convection is the fluid movement within a lava lamp.

Delta-T   Delta-T is the temperature difference between the initial (supply) fluid temperature and the final (return) fluid temperature. The delta-T can be found across any individual system component, such as a cooling coil or chiller.

Discharge   Discharge occurs in a TES tank when some, or all, of the thermal load is being served by the thermal storage tank due to the load being greater than the output of the plant.

External melt   External melt is a type of ice-on-coil thermal storage system, in which a separate heat transfer fluid is used to melt the outside of the ice than was used to initially produce the ice.

Freeboard   Freeboard is the height difference between the water surface and the shell height of the TES tank. With a steel tank, freeboard is necessary to prevent a seismic wave from hitting the tank roof during an earthquake.

Full storage   Full storage systems provide all the required cooling from the TES tank during a certain period of the day, while the cooling equipment is shut off. The cooling equipment operates during nonpeak hours, at a higher capacity than that required to meet the instantaneous cooling load, storing the surplus cooling in the TES tank.

Heat of fusion   The heat of fusion of a material or substance is the amount of heat energy required to convert one unit of mass from solid phase to liquid phase, or conversely, the amount of heat energy released or removed when one unit of mass changes from liquid phase to solid phase. The heat of fusion of water is 144 Btu/lbm, meaning that 144 Btu would be required to melt one pound mass of ice.

Internal melt   Internal melt is a type of ice-on-coil thermal storage system utilizing the same heat transfer fluid within the heat exchanger to melt the ice as was used to initially produce the ice.

Latent energy   Latent energy is the energy generated or stored as a result of the phase change of a material.

Life-cycle-cost analysis   Life-cycle-cost analysis is a process by which system costs and savings are calculated for the life of the system, taking into consideration the time value of money.

Load profile   A load profile is the set of loads for a specific period of time, used to size equipment and determine operational strategies. Peak (design) day load profiles are commonly used to size TES systems, while average day load profiles can be used for economic analysis. The load profile can be represented in both graphical and tabular forms.

Net present value   The net present value is an assessment of the current worth of a given investment. Comparison of net present values from different investments provides a ranking of cost effectiveness of each investment.

Partial storage   Partial storage systems provide a constant output from the cooling equipment for the entire day. The TES tank supplements the cooling equipment when the instantaneous cooling load is higher than the output from the cooling equipment and stores the surplus cooling when the instantaneous cooling load is lower than the cooling equipment output.

Passive thermal storage   Passive thermal storage systems do not use mechanical energy and transport to achieve thermal storage. One example of passive thermal storage systems is a high mass wall exposed to sunlight, designed to allow natural convection heat transfer to the building.

Radiation   Radiation is the form of heat transfer that occurs through the transmission of energy in electromagnetic waves. One example of radiation heat transfer is the warmth felt from the sun.

Sensible energy   Sensible energy is the energy generated or stored as a result of the temperature change in a material.

Specific heat   The specific heat of a substance is the amount of heat required to raise one unit of mass by one degree of temperature. The specific heat of water is approximately 1 Btu/lbm-°F, meaning that 1 Btu would be required to raise one pound mass of water at standard conditions.

Sustainable   At the core, sustainable implies allowing one generation to meet its needs without depriving future generations of meeting their needs. Sustainable also means that the process will not contribute to dramatic lifestyle changes required of future generations. The *Merriam-Webster online dictionary* defines *sustainable* as "1: capable of being *sustained*; 2 a: of, relating to, or being a method of harvesting or using a resource so that the resource is not depleted or permanently damaged <*sustainable* techniques> <*sustainable* agriculture> b: of or relating to a lifestyle involving the use of sustainable methods <*sustainable* society>." In engineering terms, sustainable means that the process can continue indefinitely with the proper operation and maintenance and that the economics show a good rate of return for the facility's investment. On a general basis, sustainability involves cleaning up and reusing existing sites and facilities (protecting ecosystems by using "brown field" vs. "green field"), reusing construction materials or recycled construction materials and sustainable materials (minimize resource depletion), minimizing construction waste and equipment pollution, minimizing facility water usage, minimizing facility waste streams and pollution, maximizing indoor environmental quality (which increases human productivity), and minimizing overall facility energy usage.

Thermocline   The thermocline is the zone in a stratified thermal storage tank, where there is a temperature change from the colder water that is on the bottom of the tank to the warmer water that is on the top of the tank. The thermocline occupies a volume within the tank that results in lost thermal storage capacity.

Tons of cooling   A cooling ton, or refrigeration ton, is the amount of cooling that can be achieved by melting 2000 pounds (1 ton) of ice in 1 day and is equal to 12,000 Btu/h (2000 lb times 144 Btu/lb divided by 24 hours).

Bibliography

Alter, L., "Ice Could Be Key to Storage of Renewable Energy," *Greenbuild International Conference and Expo*, Nov. 16, 2009.

A Short History of Energy Storage, District Energy Library, University of Rochester, http://www.energy.rochester.edu/storage/

ASHRAE Handbook, HVAC Systems and Equipment, Ch. 50, Thermal Storage, ASHRAE, Atlanta, GA, 2008.

Aston, A., "A Skyscraper Banking on Green: The Bank of America Tower's Environmentally Sound Design Aims for Big-Time Return on Investment," *BusinessWeek,* http://images.businessweek.com/ss/07/02/0228_greenbuilding/source/1.htm

CB&I–Steel Thermal Storage Tanks.

Cheung, K., S. Cheung, R. De Silva, M. Juvonen, R. Singh, and J. Woo, "Large-Scale Energy Storage Systems," *Imperial College of London*, 2002/2003.

Clark, L., "The Benefits of Ice-Based Thermal Energy Storage," *HPAC Engineering*, Fort Lauderdale, FL, May 6, 2010.

Clarke, E., "Thermal Storage in the US: Soon to be a given," *CSP Today*, Dec. 18, 2009.

Dincer, I., and M. A. Rosen, *Energy Storage Systems: Systems and Applications*, in Dincer, I., and M. A. Rosen, *Thermal Energy Storage*, John Wiley & Sons, 2002.

Dorgan, C. E., and James S. Elleson, *Design Guide for Cool Thermal Storage, American Society of Heating, Refrigerating and Air-Conditioning Engineers*, 1993.

Du Bois, D., "Ice Energy's 'Ice Bear' Keeps Off-Peak Kilowatts in Cold Storage to Reduce HVAC's Peak Power Costs," *Energy Priorities Magazine*, Jan. 16, 2007.

DYK Incorporated–Prestressed Concrete Tanks.

"Energy storage," McGraw-Hill Encyclopedia of Science and Technology. The McGraw-Hill Companies, Inc., 2005. Answers.com Dec. 30, 2010. http://www.answers.com/topic/energy-storage

History of Thermal Energy Storage, History, Thermal Reserve, http://www.thermalreserve.com/index.php?p=history

Holness, G. V. R., "Case Study of Combined Chilled-Water Thermal Energy Storage and Fire Protection Storage," ASHRAE Transactions: Symposia, AN-92-14-1.

Holusha, J., "On Avenue of the Americas, the Iceman Cometh," *The New York Times on the Web*, Mar. 17, 2002.

Hrbeck, R., "Cold as Ice," *Environmental Design + Construction*, Jul. 19, 2002.

Hyman, L. B., "Thermal Storage," *Orange Empire ASHRAE Chapter Meeting*, Santa Ana, CA, Apr. 27, 2010.

"Internal Melt Ice-on-Coil (IMIOC)," Cypress Ltd Shift & Save®, http://www.shift-nsave.com/pge/imioc.php

Japan Echo Inc., "Going Underground: New Thermal Storage System to Save Energy," web-japan.org, Feb. 17, 1998.

Kanellos, M., "Calmac: The Ice Age Returns to Offices," *Greentech Media*, May 6, 2009.

Katzel, J., "Take Another Look at Ice Thermal Storage Systems," *Plant Engineering*, Jul. 8, 1993.

Knebel, D., "Predicting and Evaluating the Performance of Ice Harvesting Thermal Energy Storage Systems," *ASHRAE Journal*, May 1995.

Knebel, D., "Thermal Storage with Ice Harvesting Systems," *Third Symposium on Improving Building Systems in Hot and Humid Climates*, , Arlington, TX, Nov. 18-19, 1986.

MacCracken, M., "Thermal Energy Storage Risky Myths," *ASHRAE Journal*, Sept. 2003.

Meckler, M., "Design of Integrated Fire Sprinkler Piping and Thermal Storage Systems: Benefits and Challenges," *ASHRAE Transactions: Symposia*, AN-92-14-4.

Meckler, M., and L. B. Hyman, *Sustainable On-Site CHP Systems: Design, Construction, and Operations*, McGraw-Hill, New York, 2010.

"Modular Tanks vs. Encapsulated Ice Storage," Central Iowa Power Cooperative http://cipco.apogee.net/ces/library/ctgm.asp

Mueller MaximIce® Ice Slurry Applications, Product Literature, Mueller Thermal Storage Systems, Springfield, MO, http://www.genemco.com/catalog/pdf/EELD669Fmuellericesystemliterature.pdf

Natgun Corporation–Wire Wound Concrete Tanks.

Nelson, K. P. "Ice Slurry Generator," *IDEA Conference*, Jun. 1998, San Antonio, TX.

Paksoy H. (Ed.), *Thermal Energy Storage for Sustainable Energy Consumption: Fundamentals, Case Studies and Design*, Springer, New York, 2007.

Shi, W., B. Wang, X. Li, and X. Lv, "Closed Loop External Melt Ice-on-coil Ice Storage Tank and Experimental Research on Its Charge and Discharge Performance," Tsinghua University, Beijing.

Siegenthaler, J., "Mass Management: When should a solar storage tank be online vs. offline?" *PM Engineer*, Aug. 2010.

Silvetti, B., "Application Fundamentals of Ice-Based Thermal Storage," *ASHRAE Journal*, Feb. 2002.

Silvetti, B., "Thermal Energy Storage," in *Encyclopedia of Environmental Management*, Ed. by S. Jorgensen, 2012.

Slabodkin, A. L., "Integrating an Off-Peak Cooling System with a Fire Suppression System in an Existing High-Rise Office Building," *ASHRAE Transactions: Symposia*, AN-92-14-3.

"Standard for Water Tanks for Private Fire Protection," NFPA 22, 1987 Edition.

Tarcola, A., "Fire and Ice: University of Arizona increases turbine efficiency with ice storage," *Distributed Energy: The Journal of Energy Efficiency and Reliability*, Mar.–Apr. 2009.

Thermal Energy Storage, Applications, Vogt Ice, Louisville, KY, http://www.vogtice.com/applications_thermal_energy_storage.htm

Thermal Energy Storage Benefits, Benefits, Calmac, http://www.calmac.com/benefits/

"Thermal Storage–Glycol Systems," *Central Iowa Power Cooperative*, http://cipco.apogee.net/ces/library/xctg.asp

"Thermal Energy Storage Strategies for Commercial HVAC Systems," *Pacific Gas and Electric Company*, May 1997. http://www.pge.com/includes/docs/pdfs/about/edusafety/training/pec/inforesource/thrmstor.pdf

Valenta, P., "Ice Storage," *Environmental Design + Construction*, Feb. 1, 2008.

Wang, M. J., and V. Goldstein, "Ice Slurry Based Thermal Storage Technology," *IEA-Annex 17: 7th Expert Meeting and Work Shop*, Beijing, Oct. 11-12, 2004..

Wilson, A., "Buildings on Ice: Making the Case for Thermal Energy Storage," *Environmental Building News*, 18(7), Jul. 2009.

Wilson, A., "Chill Factor," *GreenSource*, Mar./Apr. 2010.

Wilson, A., "Making Ice at Night to Cool Buildings," *BuildingGreen.com*, Jul. 4, 2009.

Index

Note: Page numbers followed by "f" indicate material in figures; by "t", in tables; by "n", in notes.

KIS0113-15 (542)
£61·99